D0031605

Astronomy

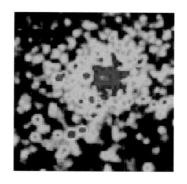

Astronomy

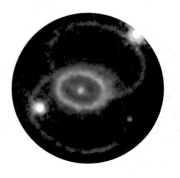

CONSULTANT EDITOR
Robert Burnham

FOG CITY PRESS

Published by Fog City Press
814 Montgomery Street
San Francisco, CA 94133 USA

Copyright © 2002 Weldon Owen Pty Ltd

Chief Executive Officer John Owen
President Terry Newell
Publisher Lynn Humphries
Managing Editor Janine Flew
Art Director Kylie Mulquin
Editorial Coordinator Tracey Gibson
Editorial Assistants Marney Richardson, Kiren Thandi
Production Manager Martha Malic-Chavez
Business Manager Emily Jahn
Vice President International Sales Stuart Laurence
European Sales Director Vanessa Mori

Project Editor John Mapps
Designer Avril Makula
Consultant Editor Robert Burnham

All rights reserved. Unauthorized reproduction,
in any manner, is prohibited.
A catalog record for this book is available from
the Library of Congress, Washington, DC.

ISBN 1 876778 81 4

Color reproduction by Colourscan Co Pte Ltd
Printed by LeeFung-Asco Printers
Printed in China

A Weldon Owen Production

Welcome to the
Home Reference Library

We have created this exciting new series of books with the help of an international team of consultants, writers, editors, photographers, and illustrators, all of whom share our common vision—the desire to convey our passion and enthusiasm for the natural world through books that are enjoyable to read, authoritative as a source of reference, and fun to collect.

Finding out about things *should* be fun. That's the basic premise of the *Home Reference Library*. So, we've ensured that every picture tells a story, every caption encapsulates a fascinating fact, and every paragraph contains useful or interesting information.

It is said that seeing is believing. We believe seeing is understanding, too. That's why in the *Home Reference Library* we have combined text and images in an imaginative, dynamic design style that conveys the excitement of finding out about the natural world. Cut-away cross-sections detail the inner workings of a termite mound or the 2,000-million-year-old rock strata of the Grand Canyon. Photographs reveal extraordinary facts about the minutest forms of animal life, aspects of the behavior of nature's fiercest predators, or the beauty of a world far beyond our planet.

Each handy-sized book is a complete source of reference on its subject. Collect all the titles in the *Home Reference Library* series to compile an invaluable encyclopedic resource that you'll return to again and again.

From the editors of the *Home Reference Library*.

Look at the stars! Look, look up at the skies!

O look at all the fire-folk sitting in the air!

The bright boroughs, the circle-citadels there!

"The Starlight Night,"
Gerard Manley Hopkins (1844–1889), English poet

CONTENTS

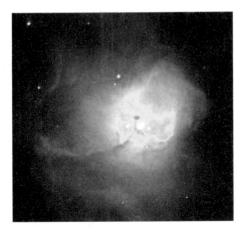

OBSERVING THE SKY 142

STAR CHARTS 186

PROBING THE UNIVERSE

*Presenting the universe in
all its majesty, mystery and
infinite variety.*

Understanding the Universe

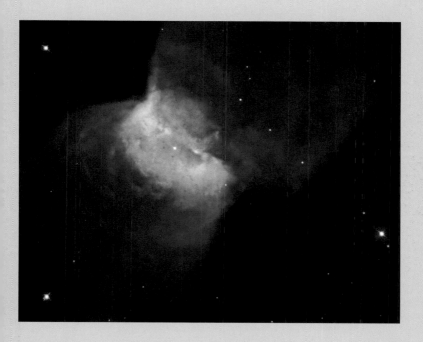

CONSTANTLY EXPANDING, the universe is of an unimaginable immensity. Our Sun is one of billions of stars in the Milky Way, which is one of billions of galaxies in the universe. And stars and galaxies are only part of the picture: There are wispy nebulas where stars are born, black holes that consume light, and perhaps even exotic invisible particles.

THE UNIVERSE

We think Earth is a big place, but trying to picture the sheer vastness of the universe is a real challenge. The Moon, our closest neighbor, is about 240,000 miles (385,000 km) from us; the Sun, 93 million miles (150 million km); and the nearest bright star, Alpha Centauri, a staggering 25 billion miles (40 billion km). The nearest large galaxy to our own, Andromeda, is about a million times farther away. The most distant galaxies yet known are more than 10 million times farther still.

THE HOME GALAXY

Along with eight other planets, Earth orbits the Sun. This Solar System is a tiny dot in the Milky Way Galaxy, a spiral galaxy of 400 billion stars that stretches 100,000 light-years from side to side.

In turn, the Milky Way is just one of billions of star-filled

OUR HOME Earth sits third in line from our star, the Sun, among the other rocky planets. Neptune and Pluto mark the outer edge of the Solar System.

OUR GALAXY The Solar System is a speck within the Milky Way Galaxy. Relatively nearby are two small galaxies, the Large and Small Magellanic Clouds.

galaxies, the building blocks of the universe. Despite its size, our Galaxy is tiny compared to the universe as a whole. Light from the nearby Andromeda Galaxy, for example, takes about 2.5 million years to reach us.

INTERSTELLAR SPACE

The Milky Way and Andromeda are the largest members of a small gathering of at least 35 galaxies called the Local Group. Our galactic family lives on the outskirts of a dense cluster of thousands of galaxies called the Local Supercluster. Astronomers believe that thousands of such superclusters are scattered throughout the universe.

What lies beyond? The Hubble Space Telescope has seen galaxies 10 billion light-years away, so far away that they appear merely as smudges near the edge of the observable universe.

OUR UNIVERSE The Milky Way (bottom left) is just one among many billions of galaxies. It is tiny compared to the universe as a whole.

MEASURING DISTANCE

Beyond our world lies a gulf in space too large to measure in miles and kilometers. Two main measurements are used instead: Astronomical Units and light-years.

A useful measure in our Solar System is the Astronomical Unit (AU), the average distance between Earth and the Sun—about 93 million miles (150 million km).

For measuring distances beyond the Solar System, we use the light-year. This is defined as the distance light travels in a vacuum in a year. A light-year is about 6 trillion miles (10 trillion km).

The Big Bang

■ In 1924, the U.S. astronomer Edwin Hubble turned the astronomical world on its ear when he announced that galaxies appeared to be moving away from Earth. What is more, Hubble discovered that the more distant the galaxy, the faster it was receding. The universe appeared to be blowing up, "inflating," like a balloon. The implications were truly astonishing: For the universe to be expanding, a force of incredible magnitude must have set matter on its outward-flowing course. The name astronomers eventually coined for this violent genesis was the Big Bang.

IN THE BEGINNING

Unfortunately, the term "Big Bang" is a misnomer that has led to some confusion. There was no "bang," no explosion in the strictest definition of the word, where one thing erupts into the space of something else. This is because there was no space. Instead, the Big Bang was more an unfolding of space and matter—from a point no larger than the period at the end of this sentence. This happened between 13 billion and 15 billion years ago.

After the bang A fraction of a second after the Big Bang, the

CREATION STORY The universe began with a bang, and in one-millionth of a second became a mass of radiation and particles. These particles formed the first atoms of hydrogen and helium. After several billion years, gravity caused clouds of gas to collapse, forming stars and galaxies.

universe was a hot, seething mass of radiation and exotic particles. This expanded and cooled, and more familiar sorts of particles formed, including the neutrons, electrons, and protons that make up everyday matter.

Matter forms Gradually, elements came into being—mostly hydrogen and helium—and these eventually collapsed under the influence of gravity to create galaxies, stars, and planets.

EVIDENCE FOR THE BIG BANG

The most convincing single piece of evidence for the Big Bang is the cosmic background radiation (CBR), a uniformly distributed all-sky glow. It is thought to be the last vestige of heat from the Big Bang itself. That initial outburst of fiery energy has cooled greatly, to −455°F (−270°C), just barely above absolute zero and some hundred million times cooler than a typical birthday candle.

The Future of the Universe

■ The universe has been expanding since the Big Bang, but will it continue to do so forever? Astronomers have come up with a number of intriguing scenarios, ranging from never-ending expansion to a long return journey that ends with a Big Crunch. In all cases, whether expansion will change at some point in the future depends on the density of the universe, and that depends on the amount of matter it contains.

POSSIBLE FUTURES

If the density of the universe is greater than a certain critical value—that is, the universe contains a sufficient amount of matter—the force of gravity will at some point bring expansion to a halt and the universe will begin to collapse, culminating in a Big Crunch. Such a universe is called "closed." If the density of the universe is less than the critical value—there is not enough matter—expansion will continue indefinitely. This is termed an "open" universe.

A flat universe There is another option. Much observational evidence suggests that the universe is "flat," precisely balanced between open and closed. That means the universe will expand forever, always decelerating, but never quite coming to a halt.

An accelerating universe Yet another possibility is that expansion may actually accelerate. Supernovas observed in distant galaxies in 1998 showed that they are 20 percent fainter than would be expected in a flat or open universe. This indicates that over the past few billion years, the expansion of the universe has sped up and carried the stars to greater distances from Earth.

A LIKELY STORY

Which possible future is most likely? Astronomers are in general agreement that the universe will continue to expand; the evidence does not favor the closed universe scenario. Beyond that, opinions are very much divided. A vigorous debate continues.

BIG CRUNCH The Big Crunch is one possible fate of the universe. At the moment (bottom left sphere), the universe is expanding. If there is enough matter in the universe, the expansion halts and switches into reverse. The end comes with an all-destroying crunch (top right sphere).

INDEFINITE EXPANSION If the universe keeps expanding from the present (bottom left sphere), galaxies will drift farther apart, stars will eventually burn out, and ordinary matter will disintegrate, leaving a boundless "sea" of elementary particles (far right sphere).

STARS

A star is a large sphere of hydrogen and helium, with a smattering of other elements, all in gaseous form. Nuclear fusion reactions in its core create enormous amounts of energy, including light and heat. No two stars are alike—brightness, surface temperature, and mass vary from one to the other.

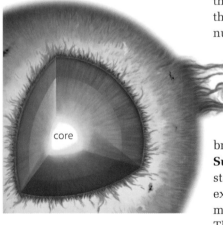

NUCLEAR FURNACE A main-sequence star, such as the Sun, has a nuclear furnace at its core. Nuclear reactions fuse hydrogen into helium and release energy. As each bit of energy reaches the surface, it flies into space—and the star shines.

STAR TYPES

Stars start their lives on what astronomers call the main sequence. These are often called dwarfs, although the term is misleading: some dwarfs are 20 times bigger than the Sun and 20,000 times brighter. Red dwarfs are the most common of all stars.

White dwarfs Smaller than red dwarfs, white dwarfs are typically the size of Earth but the mass of the Sun. They are stars whose nuclear fires have gone out.

Red giants After the main-sequence stars, the most common stars are the red giants. They have the same surface temperature as red dwarfs, but are much brighter and larger.

Supergiants These are the largest stars of all—Betelgeuse, for example, is close to 600 million miles (1 billion km) in diameter. These monsters typically have a mass similar to the Sun's, but, if they traded places with the Sun, their atmospheres would envelop the planets of the inner Solar System.

MULTIPLE STARS

Most stars have at least one companion
star: as a singleton, our Sun is in the
minority. Double stars probably form
when several different parts of the
parent cloud of dust and gas begin
to collapse at once. Sirius (above),
is a double, made up of a brilliant
bluish supergiant, the sky's
brightest star, plus a faint white
dwarf companion.

Stars with two or more
companions are known as
multiple stars.

COMPARING SIZE The small white ball
at right represents a white dwarf star.
The Sun, the yellow globe, is huge by
comparison, but the red giant behind
it is 20 to 40 times larger, and the red
supergiant is roughly 800 times larger.

The Life of a Star

■ All stars begin their lives in the same way, by condensing out of clouds of gaseous material. But the details of their lives—and deaths—differ greatly.

TWO PATHS

A star's lifespan depends on its mass. Massive stars form quickly, in, perhaps, hundreds of thousands of years compared to the Sun's tens of millions of years, but have relatively brief lives. A star that is eight times more massive than the Sun will shine for only 40 million years,

A NURSERY IN THE EAGLE Stars are formed inside clouds of gaseous material called nebulas. Light from newborn stars is eroding the gas in these columns, part of a stellar nursery in the Eagle Nebula.

while a star like the Sun can live for 10 or 11 billion years.
Stellar deaths Death, too, depends on mass. After a massive star exhausts its nuclear fuel, it dies in a blaze of glory, as a supernova (see pp. 24–5); the death of a smaller star, such as the Sun, is more subdued, as shown in the diagram below.

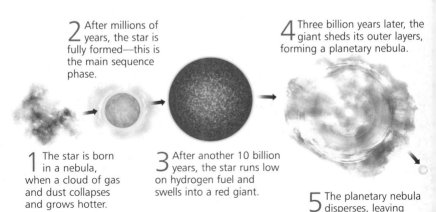

2 After millions of years, the star is fully formed—this is the main sequence phase.

4 Three billion years later, the giant sheds its outer layers, forming a planetary nebula.

1 The star is born in a nebula, when a cloud of gas and dust collapses and grows hotter.

3 After another 10 billion years, the star runs low on hydrogen fuel and swells into a red giant.

5 The planetary nebula disperses, leaving behind a white dwarf, the star's white-hot core.

Supernovas and Novas

■ Novas and supernovas are superficially similar: both involve stars that show unpredictable and significant increases in brightness. In truth, however, they are very different stellar beasts. A nova is a star that suddenly increases in brightness by as much as 10 magnitudes then declines to a value close to its original magnitude. The star is not destroyed and, indeed, may repeat the behavior thousands of years later. A supernova is more final. It is the cataclysmic death of a star.

SUPERNOVAS

Stars that are much more massive than the Sun reach the end of their lives in violent supernova explosions. A supernova is a hundred times more luminous than a nova and can outshine hundreds of billions of stars in its galaxy. These stellar fireballs are instrumental in distributing certain elements, such as iron and nickel, throughout the galaxy.

Binary stars Two types of supernova take place, depending on what kind of star is involved.

The first occurs in a double-star (binary) system. Gas from a moderately massive star accumulates onto a white dwarf, pushing the white dwarf past the point where its internal pressure can counteract gravity. Eventually, the gas causes a nuclear explosion that probably disrupts the entire star system.

Massive stars If a star is particularly massive—about eight times as massive as the Sun—and its nuclear fuel has been expended, it cannot support its own outer layers. The core first collapses then rebounds catastrophically, blowing off the star's outer layers.

Some supernovas are so bright that they can be seen with the unaided eye. The last such naked-eye supernova occurred in 1987 in the Large Magellanic Cloud.

Supernova leftovers No matter how the supernova occurred, the expanding blanket of material which formed the bulk of the star collides with the surrounding interstellar medium to produce an expanding shell of gas. This is known as a supernova remnant.

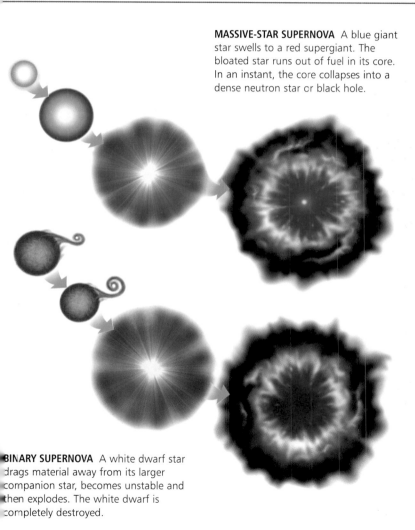

MASSIVE-STAR SUPERNOVA A blue giant star swells to a red supergiant. The bloated star runs out of fuel in its core. In an instant, the core collapses into a dense neutron star or black hole.

BINARY SUPERNOVA A white dwarf star drags material away from its larger companion star, becomes unstable and then explodes. The white dwarf is completely destroyed.

If the collapsed core of a star survives the explosion, it may remain behind in the form of a neutron star. The core becomes extremely dense and is made up entirely of neutrons—tiny particles found in the nucleus of an atom. A neutron star may be visible as a pulsar—a rapidly spinning object that flashes bursts of radio waves.

A black hole forms when an extremely massive star explodes. Too large to be a stable neutron star, the massive star keeps collapsing until it eventually disappears, leaving only a source of gravity so strong that even light waves cannot escape from it.

NOVAS

Novas occur in a binary system in which one star is a white dwarf and the other is an ordinary star. The white dwarf has greater gravity than its companion, and is able to siphon hydrogen-rich gas from it. The gas swirls into a disk around the white dwarf before spiraling down onto the dwarf's surface. When enough gas has

NOVA IN THE MAKING Below: Gas swirls down to a tiny white dwarf from its much larger companion star. The gas may trigger an explosion, causing a nova.

accumulated on the surface, the dwarf erupts in a nuclear explosion, resulting in a nova outburst.

Seen from Earth, the star system brightens by perhaps 10 magnitudes over a few days and remains bright for several days before slowly fading. The explosion may be repeated.

SUPERNOVA REMNANT Right: This is what is left of a star that went supernova 15,000 years ago—streamers of gas called the Veil Nebula or Cygnus Loop.

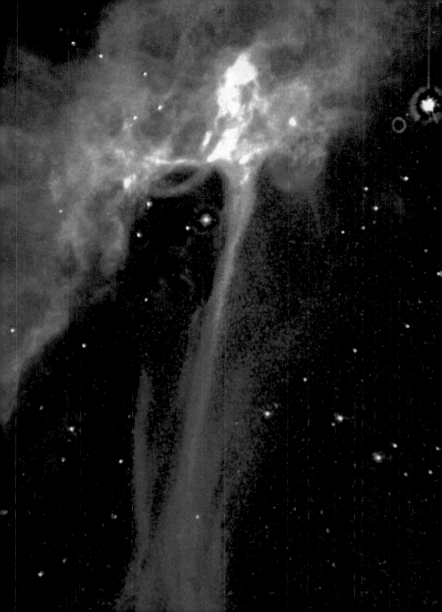

Black Holes

■ A black hole swallows up anything that comes close to it—even light cannot escape its voracious gravity.

TWO TYPES

A black hole forms when a high-mass star collapses at the end of its life. That produces an object roughly the mass of a star. But black holes can also form when a large number of stars come close together in the dense core of a galaxy. The result is a super-massive black hole. Our own Milky Way is thought to have one of these at its center.

Structure At the heart of a black hole, matter is crushed into a singularity, a point where the laws of physics cease to operate. Surrounding the singularity is an imaginary surface known as the event horizon. Surrounding the black hole itself is an accretion disk, a vast sheet of gas and dust.

LOOKING FOR HOLES

Although black holes themselves cannot be seen, the effects of their gravity can be measured. The black holes' gravity squeezes the accretion disk with great force, heating it to extremely high temperatures. Before matter is sucked into the black hole, it radiates X-rays, radio waves, and large amounts of visible energy.

Hubble's evidence We also have evidence from the Hubble Space Telescope. Photos have shown accretion disks—a telltale sign of a black hole.

DOWN A BLACK HOLE

What would happen if you fell into a black hole? Though time would seem to be passing at a normal rate, anyone watching you would see you slow down until you seemed to be in suspended animation. Space is so warped around a black hole that you would feel a much stronger gravitational pull on your feet than your head (assuming you were going in feet first), and this difference would rip you apart.

Some say that if you could survive the tidal ripping and avoid the crushing force of the singularity at the black hole's heart, you might find yourself in a different universe.

WHIRLPOOL A black hole swallows everything near it. In doing so, it creates an accretion disk, a whirlpool of material made from torn-apart stars and clouds of gas. As this material falls into the hole, it emits copious amounts of energy.

Variable Stars

Not all stars shine with a steady light. Many fluctuate in brightness over periods ranging from minutes to years.

CLASSES OF VARIABLES

Astronomers have listed some 30,000 variable stars. There are three broad classes.

Pulsating variables These brighten and fade as their outer layers rhythmically contract and expand. Two common types are Cepheid and Mira variables. Cepheids have very regular periods ranging from one to several days. Mira stars pulsate like Cepheids do, but their cycles are less regular and their periods are longer, ranging from 80 days to five years.

Cataclysmic variables These exhibit sudden and large brightness outbursts. Many are novas, binary systems that erupt once in a cycle that can last thousands of years (see page 26).

Eclipsing variables These are binary systems in which one star eclipses the other during each orbital period. From Earth, we see this as a periodic decrease in light output, followed by a return to normal brightness.

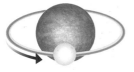

1 Bright white star blocks some light from orange star— medium brightness.

2 White star beside orange star— maximum brightness.

ECLIPSING Some double stars are aligned in such a way that one passes in front of the other, and then behind it, so that the light from the system varies.

3 Orange star blocks all light from white star—minimum brightness.

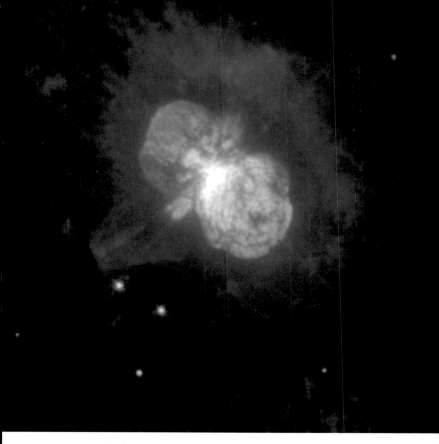

 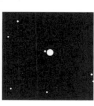

CATACLYSMIC AND PULSATING Light from a cataclysmic variable, such as Eta Carinae (above), varies erratically—there is no predictable pattern. Pulsating variables, by contrast, follow a regular pattern over a definite period of time, from minimum (far left) to maximum (left).

Star Clusters

■ Stars not only come in pairs and multiples, but also in clusters that range in size from tens of stars to more than a million. Mutual gravity holds them together, and they drift through space in the same direction and at the same speed, like a school of fish. Clusters vary in age from a few million to billions of years.

OPEN CLUSTERS

Loose assemblages of stars, containing at most a few thousand members, are called open or galactic clusters. Some 1,200 open clusters are known in the Milky Way Galaxy, and many have been observed in nearby galaxies. The smallest are just a few light-years across; the biggest may be a hundred times larger.

Open clusters are found along the spiral arms of the thin galactic plane. Most of these stars are younger than the Sun, and some are among the youngest stars we can see.

In cosmic terms, open clusters have short lives. Their stars drift apart and scatter in only a few hundred million years.

Some open clusters Examples of this type include the Pleiades in Taurus, the Double Cluster in Perseus, and the Praesepe or Beehive Cluster in Cancer.

GLOBULAR CLUSTERS

Globular clusters are denser and more tightly bound than open clusters. They also contain many more stars—some globulars have more than a million members.

HERCULES The Hercules Cluster (M 13) is a globular that contains at least a million stars within a region only 100 light-years across.

Most of the globulars we see are located in a halo surrounding our Galaxy, and similar halos of globulars have been observed around other galaxies.

Globulars are ancient. The average age of most of them is about 11.5 billion years—nearly the age of the universe itself.

PLEIADES The Pleiades is probably the best-known open cluster, plainly visible to the naked eye in Taurus. The surrounding dust is revealed clearly only in photos.

Some globular clusters Two fine globulars are Omega Centauri, in Centaurus, and the Hercules Cluster (M 13), in Hercules.

NEBULAS

A stronomers originally used the word "nebula" to describe anything that appeared blurry or cloudlike through a telescope. Many of these clouds, it turned out, had distinct spiral or oval shapes and were later recognized to be galaxies. But a large number had no particular shape and appeared to have stars embedded within them. These are true nebulas, complexes of interstellar gas and dust.

TYPES OF NEBULAS

Most nebulas are lit up by light from stars within them. The biggest nebulas do not shine at all—they can be seen only when they block the light of more distant stars.

Emission nebulas In these, atoms are excited, or ionized, by ultraviolet radiation given off by hot stars inside them. This makes these nebulas glow brightly, and shine colorfully in photos. One of the best is the Orion Nebula, in the sword of Orion. Emission nebulas are associated with both star formation and star death.

CLOUDS IN SPACE The Orion Nebula (above) is lit up by four central stars known as the Trapezium cluster. The Trifid Nebula (right) is a mixture of emission (pink) and reflection (blue) regions.

Reflection nebulas A reflection nebula does not shine on its own power but by the light of adjacent stars scattering off dust grains in the cloud. The tiny dust grains reflect blue light more effectively than red light, so these nebulas often appear blue in color.

Sometimes, both emission and reflection regions exist in the same nebula—the Trifid Nebula in Sagittarius is a good example.

Dark nebulas Dark nebulas are
clouds of gas and dust dense
enough to block the light from
background stars. They show up
as dark silhouettes (such as the
Horsehead Nebula in Orion) or
as "holes" in space (such as the
Coal Sack in Crux). Many dark
nebulas obscure regions of
newly formed stars.

PLANETARY NEBULAS

Quite different to the other clouds
of gas and dust are the planetary
nebulas. Through a telescope,
these often appear as small
disks—not unlike a planet,
hence the name.

But these expanding rings or
shells of gas have nothing to
do with planets; nor are they

associated with starbirth. They are the final stages in the life of small-mass stars (see pp. 22–3), and they glow because of radiation given off by the hot core of the star, now a white dwarf.

The Ring Nebula in the constellation Lyra and the Dumbbell Nebula in Vulpecula are two of the finest planetary nebulas in the sky.

SUPERNOVA REMNANTS

Like planetary nebulas, supernova remnants are expanding shells of gas—but they are produced by

PLANETARY NEBULA The Cat's Eye Nebula is a planetary nebula, the wreckage of a star that threw off its outer layers—in this case, about a thousand years ago.

massive stars that have exploded in a catastrophic eruption (see pp. 24–5).

Supernova remnants less than a thousand years old are strong sources of radio waves and X-rays. A prominent example is the Crab Nebula in Taurus, first observed as a "guest star" in AD 1054 by Chinese astronomers.

Supernova remnants are usually fainter and less symmetrical than planetary nebulas. As remnants age, they become even less regular in shape, appearing as huge rings, filaments, or arcs of tenuous gas.

Other prominent remnants include the Veil Nebula in Cygnus and the Vela supernova remnant.

DARK HORSE The Horsehead Nebula is a dense black cloud of dust that can be seen only because it blots out the light coming from the glowing streamers of an emission nebula

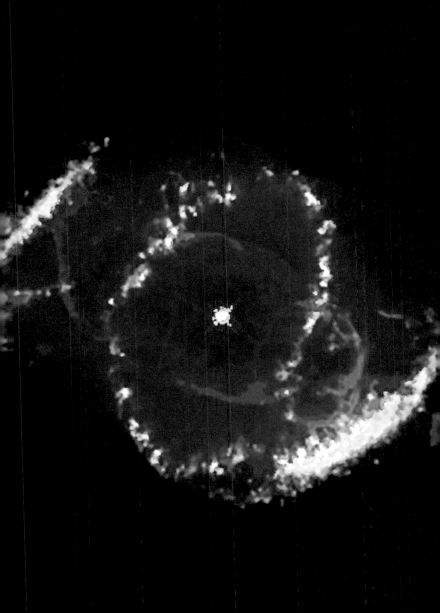

GALAXIES

Stars are not randomly scattered in space. Instead they gather in galaxies, bound together by gravity. There are billions of galaxies in the universe.

A GALAXY OF GALAXIES

Astronomers classify galaxies by shape. Galaxies come in a great variety, from pinwheels and spheres to ellipsoids and shapeless blobs.

Elliptical galaxies These are shaped like spheres, although many appear flattened or lens-shaped. The largest have diameters of 100,000 light-years or more and have a mass 100 trillion times that of the Sun. More common are the dwarf ellipticals, which may have a diameter of only 1,000 light-years or so and have a mass a few million times that of the Sun.

Spiral galaxies Most bright galaxies are spirals. These range in size from 15,000 to 150,000 light-years across and may contain anywhere between 10 billion and 10 trillion times

ELLIPTICAL GALAXY Elliptical galaxies vary in size from dwarfs to giants. M 87 is thought to have a mass equivalent to more than a million Suns.

the mass of the Sun. The arms unwind from a bright central region, called the nucleus, and wrap around the disk. A spiral galaxy's hub may be prominent in some cases and almost non-existent in others.

Barred spiral galaxies In this type, the bright stars and hot gas of the inner regions extend for thousands of light-years from either side of the center in a

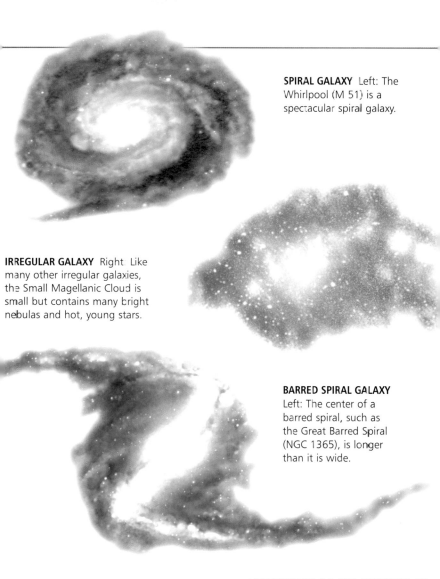

SPIRAL GALAXY Left: The Whirlpool (M 51) is a spectacular spiral galaxy.

IRREGULAR GALAXY Right Like many other irregular galaxies, the Small Magellanic Cloud is small but contains many bright nebulas and hot, young stars.

BARRED SPIRAL GALAXY Left: The center of a barred spiral, such as the Great Barred Spiral (NGC 1365), is longer than it is wide.

straight "bar" before wrapping back around the galaxy in the form of arms. In obvious cases, each arm looks like a scimitar or scythe blade.

Lenticular galaxies These have shapes that fall between the highly flattened elliptical and spiral categories. They have a central nucleus and are lens-shaped like spiral galaxies but, like the ellipticals, show little or no evidence of a spiral structure.

Irregular galaxies As their name suggests, irregular galaxies have no defined shape. They look patchy and sprawling, although some have conspicuous bars, and others have bars plus a distinctive but faint spiral arm pattern.

ACTIVE GALAXIES

Some galaxies are very turbulent places indeed. Astronomers call these active galaxies, and they are intense sources of infrared, radio, and X-ray energy. The emissions are evidence of violent activity in the galactic cores. Quasars are extremely distant objects believed to be the cores of active galaxies.

COLLIDING GALAXIES

Sometimes two or more galaxies pass very close to one another. When they "collide," few stars are actually hit because they lie too far apart. But the galaxies are pulled out of shape and may eventually merge. Also, the gravitational pull of the collision triggers starbirth.

COLLISION IN PROGRESS The Hubble Space Telescope captured a collision between the galaxies NGC 2207 and IC 2163 (left). They may well merge into a single galaxy billions of years from now.

QUASAR The bright cores of active galaxies that vary in brightness over weeks or days are known as quasars. These objects are incredibly distant— about 1.5 billion light-years from Earth.

Our Galaxy

■ The Milky Way is our home galaxy. Spiral in shape, it contains an estimated 400 billion stars, not to mention dense clouds of dust and gas.

SIZE AND STRUCTURE

The Milky Way is vast—the Sun is no more than a speck in one of its spiral arms. Its disk is 1,500 light-years thick, with spiral arms uncoiling to a distance of 75,000 light-years from the center. Surrounding the disk is a halo of old stars and star clusters that stretches perhaps another 75,000 light-years. Each star and nebula orbits the Galaxy's center more or less independently. Our Sun completes one orbit in about 240 million years.

Spiral arms

Astronomers have traced nearby sections of the Milky Way's spiral arms, but more distant sections are more difficult to make out, because dust gets in the way. For that reason, we are unsure about how many spiral arms there are. There may be as many as five.

The center Astronomers think that an object with a mass about 2.6 million times that of the Sun lies at the center of our Galaxy. They believe it to be a super-massive black hole.

FACE-ON Seen from above, the Milky Way looks like a pinwheel and has at least two spiral arms trailing from the disk.

EDGE-ON Seen edge-on (below), the Milky Way has a fairly thin disk, with a bulge at the center. Surrounding it is a halo of old stars and star clusters.

INSIDE VIEW From Earth, the Milky Way appears like a gossamer band of light or a high, thin cloud. In fact, it is a galaxy—our Galaxy—seen from the inside.

The Local Group

■ Like stars, galaxies tend to congregate into clusters. The Milky Way belongs to a small cluster of galaxies known as the Local Group. Astronomers know of at least 35 Local Group galaxies spread across roughly 8 million light-years of space. All member galaxies are linked to one another by the pull of gravity.

WITHIN THE GROUP
Most of the Local Group's members are dwarf elliptical and irregular galaxies containing only about a million stars each. Two large galaxies dominate—the Milky Way and the Andromeda Galaxy—and each has attracted a collection of smaller galaxies. For example, the Milky Way's satellite

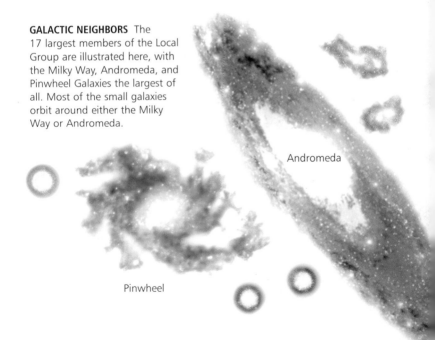

GALACTIC NEIGHBORS The 17 largest members of the Local Group are illustrated here, with the Milky Way, Andromeda, and Pinwheel Galaxies the largest of all. Most of the small galaxies orbit around either the Milky Way or Andromeda.

Andromeda

Pinwheel

galaxies are the Large and Small Magellanic Clouds and several dwarf galaxies.

OTHER GROUPINGS

Galaxy surveys have found nearly 3,000 clusters within 4 billion light-years. Compared to some of these, the Local Group is tiny. The Virgo cluster has more than 2,000 members, while the Hercules cluster contains tens of thousands of galaxies.

Superclusters Clusters of galaxies themselves belong to larger groupings called superclusters. With many other clusters, the Local Group and Virgo cluster form the Local Supercluster, some 60 million light-years across.

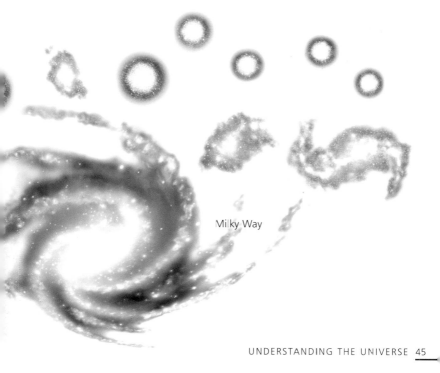

Milky Way

THE BIG PICTURE

O n a large scale, the universe is a strange place. Astronomers have found that galaxies are clumped together to form sheets and filaments separated by seemingly empty space. Even stranger, we now think that most of the matter in the universe cannot be seen.

FINDING A PATTERN

Superclusters of galaxies are not distributed evenly in the universe. They form enormous honeycomb-like structures, with vast voids between them. One

WALLS OF GALAXIES This map of superclusters shows that on a very large scale, the universe has a structure of "walls" of galaxies separated by dark voids.

ANCIENT GALAXIES This Hubble image shows the oldest galaxies ever seen. Astronomers study these ancient galaxies to see how today's galaxies evolved.

explanation for this pattern looks back 300,000 to 500,000 years after the Big Bang, when some parts of space became denser than others. The denser regions later became galaxies; the other regions became the voids between them.

DARK MATTER

The way galaxies move in clusters and superclusters shows that they are being tugged by something invisible. Most galaxies are probably surrounded by halos of this invisible "dark matter." In fact it could make up some 90 percent of the universe's mass.

Two theories The two main hypothetical candidates for this material are WIMPs (weakly interacting massive particles) and MACHOs (massive compact halo objects). Neither type has yet been directly observed, but the quest continues.

Our Solar System

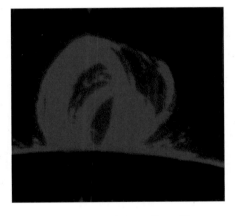

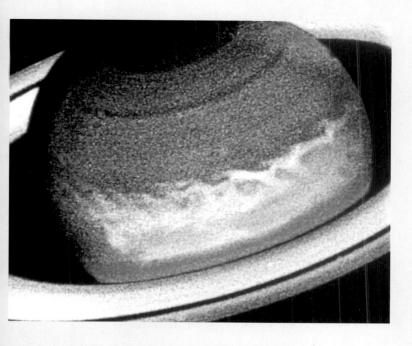

OUR BEAUTIFUL FRACTION of the cosmos is the Solar
System, dominated by the Sun and its diverse retinue
of nine planets. Large or small, the Solar System's members
are highly individual, from Mercury and its Sun-baked
craters to the frigid surface of Pluto.

THE SUN AND ITS FAMILY

The Solar System consists of the Sun; nine planets and their dozens of satellites; millions of asteroids, or minor planets; innumerable meteors and comets; and a plane of dust that pervades all of interplanetary space.

THE SOLAR NEBULA

How did all these very different objects come into being? Most astronomers believe that the Solar System formed less than five billion years ago from a cloud of hot hydrogen and helium gas and dust known as the solar nebula.

Formed from a cloud Some 4.6 billion years ago, a large cloud of cold dust and gas was drifting around the center of the Milky Way. The cloud began to collapse, possibly set off by the explosion of a nearby star.

Eventually a forerunner to our Sun was born at its center. The grains of material from the nebula consolidated into solid lumps of material. These collided and coalesced with one another to form larger bodies, which became the planets we see today.

Among the objects that did not become planets or moons are comets, small frozen objects which normally inhabit the Solar System's outer edges; asteroids, most of which orbit within the region between Mars and Jupiter; and meteoroids, fragments of comets and asteroids.

THE NATURE OF THE PLANETS

Because temperature decreases with distance from the Sun, the composition of the planets differs according to where they formed.

Inner planets The inner solar nebula was too hot for such substances as water and methane to exist as solids, so the inner planets formed from materials such as iron and silicates.

Outer planets The cold of the outer solar nebula allowed planets there to hold onto large amounts of water, ice, and other elements easily destroyed by heat. The greater mass of these planets also meant that they could sweep up large quantities of hydrogen and helium, which created voluminous atmospheres.

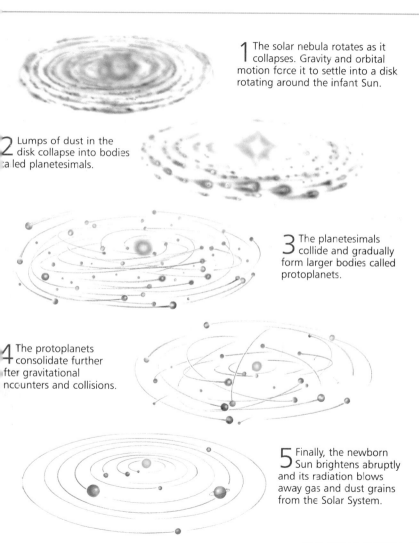

1 The solar nebula rotates as it collapses. Gravity and orbital motion force it to settle into a disk rotating around the infant Sun.

2 Lumps of dust in the disk collapse into bodies called planetesimals.

3 The planetesimals collide and gradually form larger bodies called protoplanets.

4 The protoplanets consolidate further after gravitational encounters and collisions.

5 Finally, the newborn Sun brightens abruptly and its radiation blows away gas and dust grains from the Solar System.

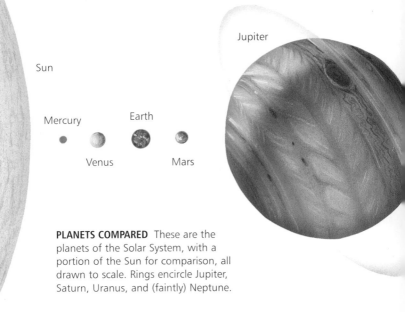

Sun

Mercury

Venus

Earth

Mars

Jupiter

PLANETS COMPARED These are the planets of the Solar System, with a portion of the Sun for comparison, all drawn to scale. Rings encircle Jupiter, Saturn, Uranus, and (faintly) Neptune.

THE SUN AT THE CENTER

All planetary bodies, from Jupiter to the smallest particle of dust, move about the Sun. The orbits are not circular but elliptical. This means that each body goes through two extreme points in its orbit, one nearest the Sun, called perihelion, and one farthest from the Sun, aphelion.

With the exception of comets, all bodies orbit in the same direction as Earth. If you could stand well above Earth's North Pole and look down on the Solar System, you would see the planets moving counterclockwise Most of the planets also rotate in the same direction as their orbital motion.

The planets go around the Sun nearly in the same plane—that is, on the same "level." Pluto is the odd man out, its orbit taking it a long way above and below the the other planets. Comets are able to approach the Sun from any direction and angle.

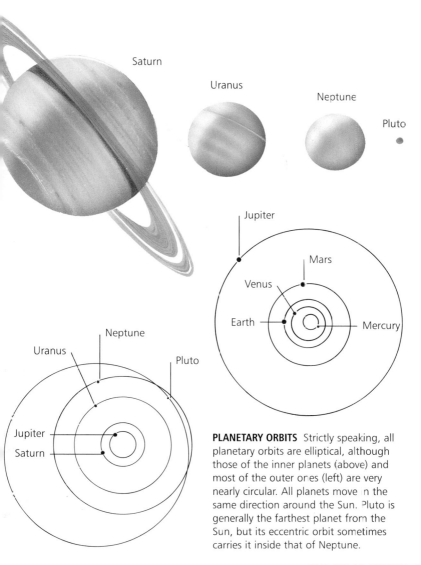

Saturn

Uranus

Neptune

Pluto

Jupiter

Mars

Venus

Earth

Mercury

Neptune

Uranus

Pluto

Jupiter

Saturn

PLANETARY ORBITS Strictly speaking, all planetary orbits are elliptical, although those of the inner planets (above) and most of the outer ones (left) are very nearly circular. All planets move in the same direction around the Sun. Pluto is generally the farthest planet from the Sun, but its eccentric orbit sometimes carries it inside that of Neptune.

Terrestrial and Gas-giant Planets

■ Most planets fall into two groups based on size, density, and chemical makeup. Pluto is neither terrestrial nor jovian, but resembles instead the rock–ice moons of the outer planets.

TERRESTRIALS

Mercury, Venus, Earth, and Mars are the terrestrial planets. These are small bodies made up mainly of rock and metals, with densities three to five times that of water. Terrestrial planets have relatively thin atmospheres.

GAS GIANTS

Jupiter, Saturn, Uranus, and Neptune are the jovian or gas-giant planets. They are all more than a dozen times more massive than Earth. In fact, Jupiter out-weighs our planet by hundreds of times.

Each jovian planet is thought to have a tiny rocky core buried under layers of hydrogen and helium thousands of miles deep. A gas-giant planet has a density near that of water, and Saturn's is actually lighter than ice.

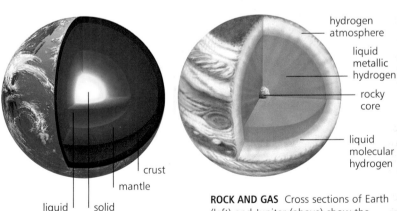

hydrogen atmosphere

liquid metallic hydrogen

rocky core

liquid molecular hydrogen

crust

mantle

liquid outer core

solid inner core

ROCK AND GAS Cross sections of Earth (left) and Jupiter (above) show the contrasting internal structures of terrestrial and gas-giant planets.

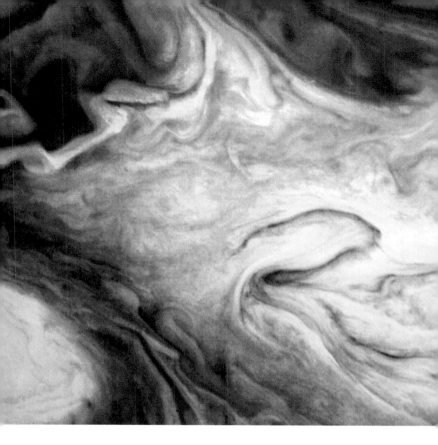

GAS BALL The photo (above) shows a close-up of Jupiter's atmosphere. What you are seeing is merely the outermost layer of a huge, mostly gaseous ball.

ROCKY WORLD From the Himalayas (here seen from space) to the core, Earth is made up mainly of dense, rocky materials such as basalt, and metals, such as iron.

The Sun

■ To us, the most important object in our sky is the Sun. Its energy powers Earth's climate and supports life. Yet the Sun is an ordinary star like a million others in the Milky Way Galaxy.

OUR STAR

The Sun is a huge ball of hot gas, mostly hydrogen and helium. In its core, high temperatures and pressures fuse hydrogen into helium, releasing energy.

Structure Fusion occurs from the center of the Sun out to perhaps a quarter of its radius. Above this is the radiative zone, in which

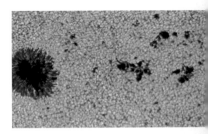

SEEING SPOTS The Sun's best-known features are sunspots, dark regions which are cooler than their surroundings.

radiation carries the energy. On top of this layer the Sun's energy moves in the same way that boiling water does. Heated from below, the gas rises to the surface, radiates energy into space, cools, then sinks again. This convective region forms the Sun's outer third.

The photosphere is the Sun's visible surface, and this is where the dark regions call sunspots form. Above this, the Sun has a complex atmosphere consisting of the chromosphere (a thin, cool layer) and the corona, which is almost as hot as the core. Both are usually only visible during solar eclipses.

FACT FILE

DISTANCE FROM EARTH
93 million miles (150 million km)
SIDEREAL REVOLUTION PERIOD
365.26 days
MASS (EARTH=1) **333,000**
RADIUS AT EQUATOR (EARTH=1)
109
APPARENT SIZE **32 arcminutes**
SIDEREAL ROTATION PERIOD (AT EQUATOR) **25.4 days**

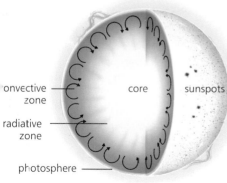

SOLAR POWERHOUSE The Sun's photosphere—the surface we can see—lies above boiling convection layers, an intermediate radiative zone, and a nuclear powerhouse at the core (left). The corona—the Sun's outer atmosphere—is heated by millions of brilliant arches, called coronal loops (above). These fountains of extremely hot gas may be up to 300,000 miles (480,000 km) high.

convective zone

radiative zone

core

sunspots

photosphere

Mercury

■ Mercury is a small, rocky, airless world with an extreme climate—roastingly hot during the day, frigidly cold at night. Its pockmarked surface is reminiscent of the Moon's.

CLIMATIC EXTREMES

Mercury's mass is just one-eighteenth that of Earth's, and much too small to retain an atmosphere. With the Sun so close, and lacking the protection from extremes that an atmosphere would give, Mercury has a very uncomfortable climate, to put it mildly. The planet's day side

MARINER 10 Most of what we know about Mercury comes from just one spacecraft, Mariner 10, which made three flybys in 1974–5. It photographed about 45 percent of the surface.

reaches 800°F (430°C) by noon. The heat dissipates into space at night, when temperatures drop dramatically, bottoming out just before dawn at –280°F (–170°C). This is the widest temperature range of any of the planets.

THE SURFACE—AND BELOW

Mercury could almost double as the Moon. The entire planet is pockmarked with impact craters ranging in size from the smallest detectable, ½ mile (1 km) across, to the giant "bull's eye" of the Caloris Basin, some 830 miles (1,340 km) wide. Highlands,

FACT FILE

DISTANCE FROM SUN **0.39 AU**
SIDEREAL REVOLUTION PERIOD (ABOUT SUN) **88.0 days**
MASS (EARTH=1) **0.055**
RADIUS AT EQUATOR (EARTH=1) **0.38**
APPARENT SIZE **5–13 arcseconds**
SIDEREAL ROTATION PERIOD (AT EQUATOR) **58.6 days**
MOONS **none**

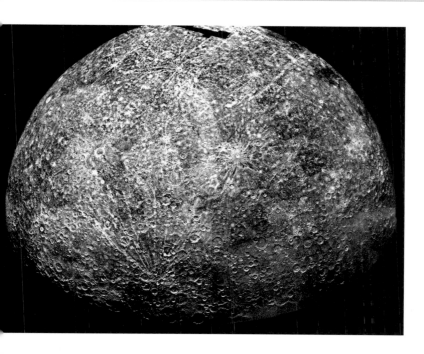

idges, and lava-flooded basins complete the picture.

nside Mercury Scientists think hat Mercury has a unique nternal structure: Like a thick ind on a piece of fruit, a crust nd mantle 330 miles (600 km) nick cover a huge metallic iron-ickel core making up 60 percent f the planet's total mass and

CRATERED SURFACE Like the Moon, the surface of Mercury is dominated by impact craters and large basins. Some craters also have prominent ray systems.

filling three-quarters of its radius. The large core may have formed that way, or it may be the result of Mercury losing some of its upper layers in a massive collision.

Venus

■ With its thick shroud of clouds, Venus kept its secrets well hidden until recent times. The Pioneer, Venera, and Magellan spacecraft have found a forbidding yet fascinating world of scorching temperatures, rocky plains, and huge volcanoes.

VENUS'S HELL

Given its greater proximity to the Sun you would expect Venus to be warmer than Earth. In fact, it is an inferno, with a surface temperature of 890°F (470°C)— hot enough to melt lead and zinc.

Greenhouse

The explanation for this lies in Venus's atmosphere of carbon dioxide, a layer so dense that the pressure at the surface of the planet is nearly a hundred times that of Earth. This thick blanket traps the Sun's heat at the planet's surface, producing an extreme greenhouse effect. Even at night, the temperature hardly drops at all. Venus can never cool down.

Cloud layers stretch upward from an altitude of 30 miles (48 km) above the surface. These are no ordinary clouds—they consist almost entirely of sulfuric acid droplets.

SURFACE FEATURES

Nearly 85 percent of Venus's surface consists of flat lava plains which resemble the maria, or "seas," of the Moon.

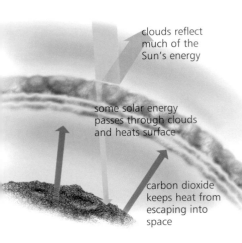

clouds reflect much of the Sun's energy

some solar energy passes through clouds and heats surface

carbon dioxide keeps heat from escaping into space

GREENHOUSE EFFECT Venus suffers from a runaway greenhouse effect. Sunlight filters through the clouds and heats the surface, but the clouds and carbon dioxide in the atmosphere keep the heat from escaping back into space.

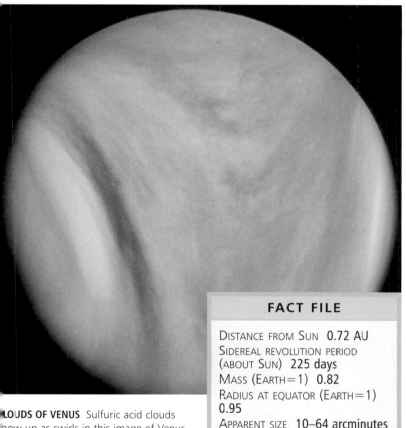

CLOUDS OF VENUS Sulfuric acid clouds show up as swirls in this image of Venus. The clouds never clear—astronomers have to use cloud-penetrating radar to pierce the haze and see details on the surface.

FACT FILE

DISTANCE FROM SUN **0.72 AU**
SIDEREAL REVOLUTION PERIOD
(ABOUT SUN) **225 days**
MASS (EARTH=1) **0.82**
RADIUS AT EQUATOR (EARTH=1)
0.95
APPARENT SIZE **10–64 arcminutes**
SIDEREAL ROTATION PERIOD (AT
EQUATOR) **243 days, backward**
MOONS **none**

Volcanoes Above the plains stand mountain ranges—including the Maxwell Mountains, which rise nearly 7.5 miles (12 km)—and thousands of volcanoes. Nearly 500 of these volcanoes are larger than about 12 miles (20 km) in diameter. Venus must once have been extremely volcanically active, but no one knows for sure if any activity occurs today.

"Continents" On a planet without oceans, it may seem odd to talk about "continents." Nevertheless, planetary scientists have identified two such landmasses. They are Aphrodite Terra, which covers a portion of the planet from the equator into the southern hemisphere, and Ishtar Terra, in the high latitudes of the northern hemisphere.

NAMING VENUS

Venus is the name of the Roman goddess of beauty and love. The International Astronomical Union (IAU) has decided that the names given to all features on the planet named after her should (appropriately enough) be female. Thus the Greek goddess Aphrodite's name has been given to a "continent" and a crater carries the name of jazz singer Billie Holiday. There is one exception, though: Scottish physicist James Clerk Maxwell is "the only man on Venus," with Maxwell Montes, the Maxwell Mountains, named after him.

Craters Venus has an estimated 900 impact craters, all bigger than about 2 miles (3 km) in diameter. It seems likely that the dense atmosphere has protected the surface from small asteroids and comets; only large objects survive a fiery descent to find their way to the ground. The craters are fairly young in

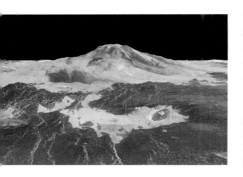

MAAT MONS This image from the Magellan spacecraft shows one of Venus's many volcanoes, Maat Mons. It rises 5 miles (8 km) above the plains.

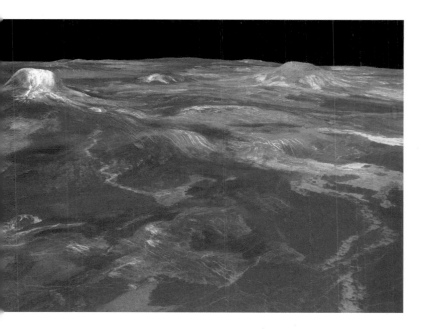

AVA PLAINS Lava plains cover most of ᵉnus's surface. This Magellan image ᵒws the Eistla region of the planet.

ᵉological terms, at less than
00 million years old. Volcanic
ᶜtivity has probably destroyed
igns of earlier impacts.

ᴸOW AND BACKWARD
ᵉnus rotates very slowly—once
ᵛery 243 Earth days, 18 days
ᵒnger than it takes to circle the

Sun. And it spins not from west
to east, like the other planets, but
from east to west.

This slow retrograde motion
has a strange effect on the
Venusian calendar. If you were
on Venus, you would see the Sun
rise in the west, cross the sky, and
set in the east some 59 Earth days
later. Perhaps an early collision
between Venus and an asteroid
or comet set the planet on its
backward course.

Earth

■ Earth is just another planet in the Solar System. True, perhaps, but the more we learn about the other planets, the more we appreciate how special the third rock from the Sun really is.

A LIVING WORLD

Earth is unique—a geologically active world cloaked in water and oxygen, with a diversity of life.
Geological activity Geologically, Earth is very active. Its surface is split into plates, floating on a rocky mantle. Earthquakes, volcanic activity, and mountain-building are concentrated along the boundaries of these plates.
Water Water is found in oceans (which cover 70 percent of Earth's surface), lakes and rivers; below the surface as groundwater; locked up in frozen polar ice caps and glaciers; and as vapor carried in the atmosphere. Earth is the only planet where temperatures allow surface water to exist in solid, liquid, and gaseous states.
Atmosphere Earth's atmosphere is rich in nitrogen and oxygen. It sustains life, protects us from the Sun's higher energy radiation, and drives our weather (by interacting with the Sun's heat).

FACT FILE

DISTANCE FROM SUN **1.00 AU**
SIDEREAL REVOLUTION PERIOD (ABOUT SUN) **365.26 days**
MASS (EARTH=1) **1.0**
RADIUS AT EQUATOR (EARTH=1) **1.0**
SIDEREAL ROTATION PERIOD (AT EQUATOR) **23.9 hours**
MOONS **1**

STORM WARNING Earth's atmosphere creates dramatic weather, such as this severe storm over the Pacific Ocean viewed from space.

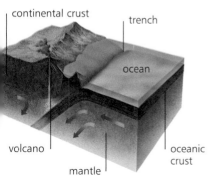

continental crust

trench

ocean

volcano

oceanic crust

mantle

THE VIEW FROM SPACE Swirling white clouds of water vapor make Earth a brilliant beacon in the Solar System.

COLLIDING PLATES Earth's crust and part of the mantle beneath it form a zone that is broken into plates. When a thin oceanic plate collides with a continental plate, the oceanic plate is drawn under and melted.

Earth's Seasons

■ The seasons happen because Earth is tilted—its rotation axis tips 23.5 degrees to its orbit.

THE TILTED EARTH

On one side of our orbit, Earth's Northern Hemisphere tilts toward the Sun. The Sun appears high in the sky, producing long, hot summer days.

Half a year later, Earth has moved to the opposite side of the Sun. Now, that same hemisphere is tilted away from the Sun. Winter days are short and the Sun appears low in the sky. The Sun's energy enters our atmosphere at a shallow angle, spreading the energy over a large surface area and diminishing its warming power.

TURNING POINTS

Four special dates punctuate the calendar. Around June 21, the North Pole is tilted most directly toward the Sun. This is the solstice, and it marks the start of summer in the Northern Hemisphere and winter in the Southern. Around December 21,

REASONS FOR SEASONS Because Earth's axis is tilted, the Sun's rays strike at various angles during the planet's orbit. The varying amounts of solar energy produce the seasonal changes.

northern winter

Solstice
21 December

southern summer

another solstice occurs when the North Pole is tilted most directly away from the Sun. This marks the start of the northern winter and the southern summer. The equinoxes, in March and September, are times when the Sun is directly over the Equator and day and night are equally long everywhere on Earth.

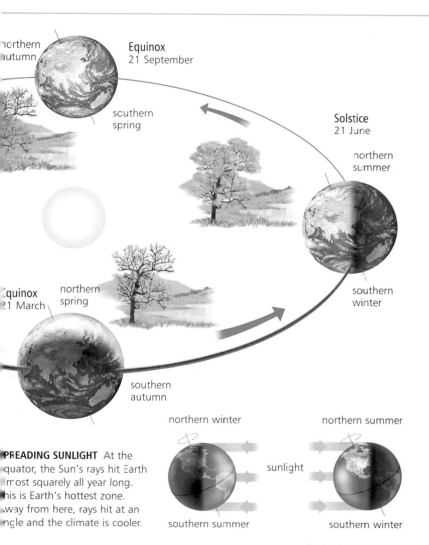

Equinox
21 September

northern
autumn

southern
spring

Solstice
21 June

northern
summer

southern
winter

Equinox
21 March

northern
spring

southern
autumn

SPREADING SUNLIGHT At the
equator, the Sun's rays hit Earth
most squarely all year long.
This is Earth's hottest zone.
Away from here, rays hit at an
angle and the climate is cooler.

northern winter

sunlight

southern summer

northern summer

southern winter

The Moon

■ The Moon is the first celestial object most new telescope owners look at—and with good reason. The sight of craters, rays, and mountains is spellbinding.

A VIOLENT PAST

The favored theory as to how the Moon came into being is that about 4.5 billion years ago a huge object struck Earth, melting the object, plus most of Earth, and sending a spray of rock into space. The spray cooled into a ring of rocky debris, which eventually coalesced and solidified into the Moon.

FACT FILE

DISTANCE FROM EARTH
239,000 miles (384,000 km)
SIDEREAL REVOLUTION PERIOD
(ABOUT EARTH) **27.3 days**
MASS (EARTH=1) **0.012**
RADIUS AT EQUATOR (EARTH=1)
0.272
APPARENT SIZE **31 arcminutes**
SIDEREAL ROTATION PERIOD (AT EQUATOR) **27.3 days**

Bombardment For 500 million years after its formation, the Moon was bombarded by asteroids and meteorites. The biggest impacts created basins hundreds of miles across that later flooded with lava. These dark regions are known as maria (singular, mare) or "seas." The remainder of the Moon's surface, unaffected by lava flooding, forms the bright and intensely cratered highlands. Relatively recent impacts are responsible for the bright streaks called rays.

Today, the Moon is quiet, impacts being rare and volcanic activity at least a billion years in the past.

THE FAR SIDE

The Moon completes one orbit around Earth in 27.3 days, which also happens to be how long it takes to complete one rotation on its axis. This means that it keeps one side turned perpetually toward Earth. We cannot see the far side, but images taken by spacecraft show a battered surface similar to the near side.

FACE OF THE MOON The Moon's main surface features can easily be seen by the unaided eye (above). The dark patches are maria, lava-filled basins mistaken for dried-up seas by early observers. The bright areas are highlands, where most craters are located. Craters range from tiny pits to large bowls with central mountain peaks and walled plairs (left).

Phases of the Moon

■ For thousands of years, the changing face of the Moon has fascinated skywatchers. In fact, the earliest known astronomical record may be a 32,000-year-old bone marked with what could be the Moon's phases.

THE LUNAR CYCLE

Earth's rotation on its axis gives us the day; the Moon's motion around Earth gives us the month. The Moon takes 27.3 days to return to the same spot in the stars, a period known as the sidereal month. A more obvious cycle is the time it takes the Moon to go through a complete cycle of phases.

Phases Just like Earth, half the Moon is always illuminated by the Sun and half is always dark. Contrary to widespread belief there is no perpetual "dark side of the Moon" (although the Moon's far side cannot be seen from Earth, see p.68).

The Moon presents a varying amount of its lit face to us during its monthly orbit, because of the changing relationship between it, the Sun, and Earth. These are known as its phases. At Full Moon, it is opposite the Sun from our point of view, so we see the entire side of it that faces us lit with sunlight. At New phase, the Sun shines on the far side of the Moon; the sunlit face is hidden. At other phases, we see only part of the Moon's sunlit surface. All the time, whether it is lit or not, we are only ever looking at the one face of the Moon.

THE TIDES

Anyone who has visited an ocean beach will be aware of the rise and fall of the tides. The Moon's gravity, and to a lesser extent that of the Sun, causes two high tides on Earth each day—one every 12 hours and 25 minutes. The gravitational pull of the Moon tugs Earth, causing the waters on the side facing it to pile up, accounting for one high tide. The high tide on the other side of Earth arises because the Moon's gravity is such that Earth itself is pulled a little toward the Moon. This results in the waters on the far side being "left behind" and piling up.

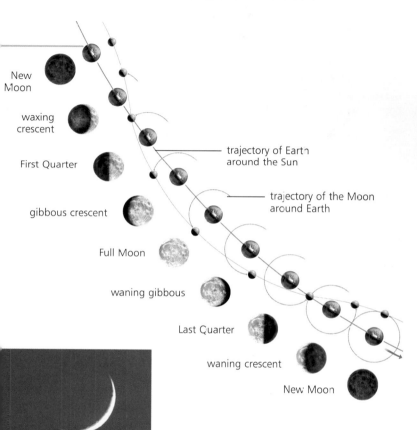

New Moon

waxing crescent

First Quarter

gibbous crescent

Full Moon

waning gibbous

Last Quarter

waning crescent

New Moon

trajectory of Earth around the Sun

trajectory of the Moon around Earth

WAXING AND WANING Throughout the course of a month, we see different amounts of the Moon lit up, depending on where it is in its orbit around Earth. In the phases between New and Full, the Moon seems to be growing larger (waxing) in the sky. Between the Full Moon and the next New Moon, it seems to be growing smaller (waning).

HANGING FACE The Moon's hanging face led many early eoples to associate it with nyths of creation and rebirth.

Mars

Mars is often called the most Earth-like planet. It has basins, plains, and highland regions that are recognized as continents. It has four seasons and its day lasts about the same as ours. Like Earth, Mars has polar ice caps, and it retains an atmosphere.

MOONS Mars's two moons are tiny, and may be captured asteroids. The larger, Phobos, is 8½ miles (13.5 km) in diameter; Deimos is only 7½ miles (12 km) across.

AN UNEARTHLY WORLD

A close look at the Red Planet shows that these similarities are only skin deep.

Climate The thin Martian atmosphere—95 percent carbon dioxide—offers only a small barrier to escaping heat. Temperatures range from −193°F (−125°C) in the polar winter to 62°F (17°C) in the southern summer. At some places on Mars, it is cold enough for carbon dioxide to freeze out of the atmosphere as dry ice.

Red desert The surface of Mars is dry and desolate, with no ecosystem or oceans. The planet's ruddy color—detectable even by the naked eye—comes from its rusty rocks and dust. A telescope shows an ocher-colored surface with darker markings—vast lava flows and boulder fields. Huge dust storms rack the planet.

VANISHED OCEANS?

Four and a half billion years ago, Mars was a very different planet indeed. It had a thicker atmosphere and, judging from the

---

FACT FILE

DISTANCE FROM SUN **1.52 AU**
SIDEREAL REVOLUTION PERIOD
(ABOUT SUN) **687 days**
MASS (EARTH=1) **0.11**
RADIUS AT EQUATOR (EARTH=1)
0.53
APPARENT SIZE **4–25 arcseconds**
SIDEREAL ROTATION PERIOD (AT
EQUATOR) **24.6 hours**
MOONS **2**

many rivulets, channels, and canyons that have been seen by space probes, abundant water. Other spacecraft findings point to thick deposits of sedimentary rocks, possibly laid down in now-vanished lakes or oceans.

Where is the water now? Some water still exists in the planet's

FACE OF MARS From space, Mars is a dry and desolate-looking place. Across its center stretches the Valles Marineris canyon, 2,500 miles (4,000 km) long.

atmosphere and polar caps. There may also be water beneath the ground at depths of 330 to 1,300 feet (100 to 400 m).

VOLCANOES AND CANYONS

Mars is not all rock-strewn plains and channels. It has some spectacular features.

Olympic majesty One of the most prominent is Olympus Mons, perhaps the largest volcano in the Solar System. It rises some 13 miles (21 km) above the surrounding plains on the planet's western hemisphere. Three other large volcanoes lie to the southeast in a region of ancient volcanic activity called the Tharsis bulge, or ridge.

OLYMPUS MONS Olympus Mons is huge: 340 miles (550 km) across and 13 miles (21 km) high. Its eruptions and lava flows must have been massive.

Grandest canyon The Valles Marineris, just south of the equator, is also remarkable. This system of canyons up to 4 miles (7 km) deep forms an immense gash stretching some 2,500 miles (4,000 km) across the planet. Scientists think that activity in the Tharsis bulge broke open the crust in this region and widened the canyon as ice washed out of the canyon walls.

LIFE ON MARS?

Mars is well embedded in the popular imagination as the most likely abode of life outside Earth. We now have evidence of land-forms caused by flowing water. That raises the possibility that life may have existed when Mars's climate was less severe. The 1976 Viking missions, however, found no traces of organic compounds in the soil. Whether life once existed on Mars—or if it still exists—is a question that only extensive exploration will settle.

VIKING The Viking 2 lander reached Mars in 1976, and began looking for traces of life. The results, and those of Viking 1 elsewhere on the planet, were negative.

Asteroids

Scattered throughout the Solar System are many thousands of bodies called asteroids.

ASTEROID TYPES

Asteroids, also known as the minor planets, are metallic, rocky objects without atmospheres. They consist of material that failed to form a planet-size body when the Solar System was taking shape. More than 35,000 asteroids have well-surveyed orbits, but establishing exactly how many exist is probably impossible.

Main belt asteroids Most asteroids with known orbits occupy a vast doughnut-shaped ring between Mars and Jupiter called the main asteroid belt. Sixteen of these asteroids have diameters of 150 miles (240 km) or greater. Most asteroids in the main belt take between three and six years to complete a full circuit around the Sun.

Trojan asteroids Jupiter has two clutches of asteroids traveling with it in its orbit, one ahead and one behind. They are known as the Trojan asteroids, and are traditionally named after figures from the Trojan Wars.

Near-Earth asteroids Other asteroids have trajectories that take them toward us. An asteroid that comes within 121 million miles (195 million km) of the Sun is known as a near-Earth asteroid (NEA). Some of these have orbits that cross Earth's, with a risk of collision. There are about 500 known NEAs, but astronomers think there may be thousands more that are large enough—1/2 mile (1 km) in diameter—to cause devastation if they hit us.

IDA AND DACTYL Ida is a main belt asteroid more than 32 miles (52 km) long. The dot at right in the photo is Ida's tiny moon, Dactyl.

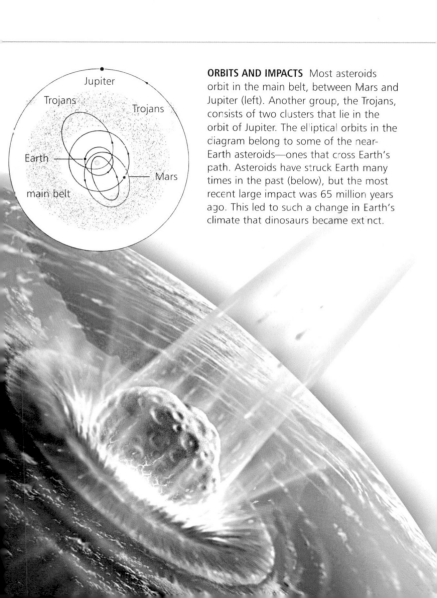

ORBITS AND IMPACTS Most asteroids orbit in the main belt, between Mars and Jupiter (left). Another group, the Trojans, consists of two clusters that lie in the orbit of Jupiter. The elliptical orbits in the diagram belong to some of the near-Earth asteroids—ones that cross Earth's path. Asteroids have struck Earth many times in the past (below), but the most recent large impact was 65 million years ago. This led to such a change in Earth's climate that dinosaurs became extinct.

Jupiter

Trojans

Trojans

Earth

Mars

main belt

Meteors and Meteorites

■ Earth experiences a continuous rain of objects, called meteoroids, which range in size from microscopic particles to small boulders. When they enter the atmosphere —visible as a streak of light, a "shooting star"—they are called meteors. Those that make it to the ground are meteorites.

VISITORS FROM SPACE

Meteoroids are fragments of asteroids or comets. Debris from comets is fragile and burns up high in the atmosphere. But a piece of an asteroid is often larger and tougher; many survive to land as meteorites.

Meteors appear 50 to 75 miles (80 to 120 km) above Earth's surface and move at speeds of between 25,000 and 160,000 miles per hour (11 and 72 km/s).
Composition Stony meteorites consist of silicates—rocky material. Iron meteorites, the second most common type, consist of 90 percent iron and 10 percent nickel with traces of silicates. The rarest meteorites are a stony-iron mixture.

STREAK OF LIGHT A meteor, or shooting star, provides a burst of movement in an otherwise still, starry sky.

METEOR SHOWERS

On most moonless nights, you could expect to see around four or five meteors per hour. Meteor showers offer something much more spectacular.
Comet debris Meteor showers occur at certain times of the year when Earth sweeps through the trail of dust from a comet. Sometimes thousands of "shooting stars" literally light up the sky— in November 1833, people on North America's east coast saw up to 200,000 meteors per hour.
Major showers Meteor showers are named after the constellation or star nearest to the point in the sky from which they appear to come, a region called the radiant. The shower that appears to come from Gemini every 14 December is called the Geminids. Other annual showers include the Perseids of August, the Leonids of November, the Quadrantids of January, and the Lyrids of April.

Iron meteorite

Stony-iron meteorite

Stony meteorite

Jupiter

■ Jupiter was named after the chief god in Roman mythology, and rightly so. Its size alone makes it king of the planets: 88,846 miles (142,984 km) across at the equator, about 11 times the figure for Earth. It is also more than 300 times as massive as Earth and twice as massive as all the other planets combined.

INSIDE A GAS GIANT

The colorful clouds of Jupiter's visible surface are just a thin layer masking the bulk of an immense atmosphere thousands of miles deep.

FACT FILE

DISTANCE FROM SUN **5.20 AU**
SIDEREAL REVOLUTION PERIOD
(ABOUT SUN) **11.9 years**
MASS (EARTH=1) **318**
RADIUS AT EQUATOR (EARTH=1)
11.2
APPARENT SIZE **31–48 arcseconds**
SIDEREAL ROTATION PERIOD (AT
EQUATOR) **9.9 hours**
MOONS **28**

Clouds Crossing Jupiter's disk are bright and dark clouds, known as zones (the light-colored bands) and belts (the dark ones). Here, wild winds blow at 400 miles per hour (640 km/h).

The Great Red Spot is the best-known cloud formation. It varies in size—and color intensity— and at its largest is 25,000 miles (40,000 km) long and 8,700 miles (14,000 km) wide.

Interior Hydrogen and helium make up most of the atmosphere. Beneath the clouds lies a layer of hydrogen gas. Moving toward the center, this layer turns to a liquid as temperatures and pressures increase. The planet's center is thought to contain a core of rocky material, 10 to 20 times the mass of Earth.

GRAVITY

Jupiter's gravitational attraction is second only to the Sun's. This force directs the fate of many comets, and it can send asteroids racing through the Solar System. It also governs a miniature Solar System of at least 28 moons.

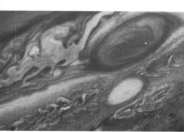

COLORFUL GIANT Jupiter presents a complex and colorful face of dark belts, light zones, and various cloud structures (above). The Jovian day spins the belts and zones into streaks and disturbances, driven by winds that blow in alternating directions parallel with the equator. The Great Red Spot (left) has been observed for more than 300 years.

The Moons of Jupiter

■ At last count, Jupiter had 28 moons, most of them fairly small. Several outer satellites are thought to be captured asteroids. The four largest—and the focus of most scientfic attention—are the Galilean satellites, discovered in 1610 by Galileo.

GALILEAN SATELLITES
The four largest moons, from Jupiter outward, are Io, Europa, Ganymede, and Callisto.
Io and Europa Io is so affected by Jupiter's tidal forces that it is in turmoil, its surface constantly being repaved by sulfurous

ICY SHELL The surface of Europa is a bright shell of ice crisscrossed by fracture lines (in red) filled with fresh ice.

volcanic eruptions. Europa, in contrast, has an icy surface resembling a planet-wide skating rink. Astronomers suspect that an ocean lies beneath it.
Ganymede and Callisto The Solar System's largest moon, Ganymede has a complex surface and perhaps a subsurface ocean like Europa's. Callisto is densely cratered, with one crater measuring some 900 miles (1,500 km) in diameter.

Io
Io is the most volcanically active body in the Solar System. Eruptions occur constantly.

Europa
Europa's surface is almost pure water-ice. Does it cover an ocean which might harbor life?

Ganymede
Ganymede is bigger than Mercury. Its surface is scarred with craters.

Callisto
This is the Solar System's most heavily cratered moon. It is mad of rock and ice.

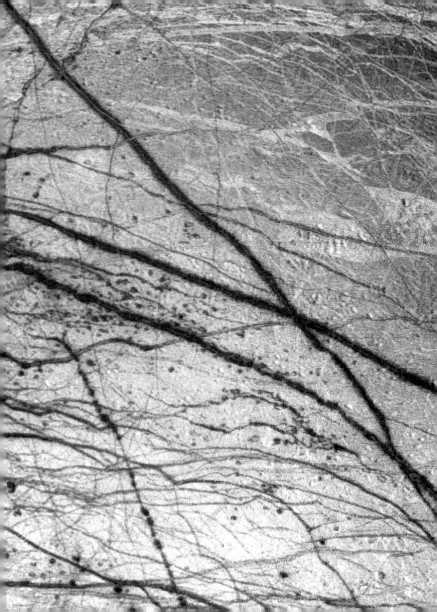

Saturn

■ Saturn is best known for its spectacular rings (see pp. 86–7), but the planet has many other wonders worthy of attention.

ABOUT SATURN

Saturn has a smaller diameter (75,000 miles or 120,660 km) than its fellow gas giant, Jupiter, and it is much lighter, with a mass equal to 95 Earths. Its composition closely parallels that of Jupiter: 74 percent hydrogen, 24 percent helium, and small amounts of methane, ethane, and ammonia Chemical reactions by the latter three cause Saturn's tan color and faint cloud banding.

Clouds The surface we see is crossed by cloud bands. Since Saturn orbits farther from the Sun than Jupiter, its environment is colder. This means it has less "weather" and so displays fewer features in its cloud tops.

INNER STRUCTURE

Scientists think that Saturn's inner structure resembles that of Jupiter. A layer of clouds covers a thick layer of fluid hydrogen that

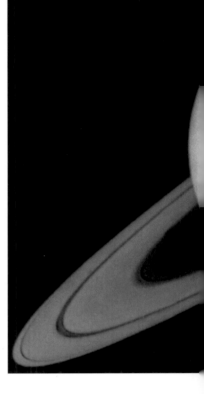

grows hotter and denser the farther it is from the surface. This probably becomes metallic about 20,000 miles (30,000 km) down. The core is thought to be a molten rocky ball weighing a dozen or more Earth masses.

FACT FILE

DISTANCE FROM SUN **9.54 AU**
SIDEREAL REVOLUTION PERIOD
(ABOUT SUN) **29.5 years**
MASS (EARTH=1) **95.2**
RADIUS AT EQUATOR (EARTH=1)
9.5
APPARENT SIZE (PLANET'S DISK)
15–21 arcseconds
SIDEREAL ROTATION PERIOD (AT
EQUATOR) **10.7 hours**
MOONS **30**

FLATTENED PLANET The Voyager 2 spacecraft snapped this close-up of Saturn
during its flyby in July 1981. Note the
obvious flattening of the planet's disk
at the poles—a direct result of Saturn's
extremely rapid rotation (just over
10 hours at the equator).

The Rings of Saturn

■ Jupiter, Uranus, and Neptune also have rings, but Saturn's are by far the most magnificent. What exactly are the rings and how did they get there?

RING STRUCTURE

The rings span 170,000 miles (270,000 km) and are no more than a few hundred yards thick. **The main rings** From Earth, we can see what look like three rings, A, B, and C. A gap, known as the Cassini division, separates rings A and B. Spacecraft images have revealed four additional rings.

RINGS AND RINGLETS This Voyager 2 image shows numerous bright and dark ringlets within the C ring. Each ringlet contains myriad icy fragments and boulder-size rocks.

Composition The rings are made up of thousands of ringlets, each consisting of ice and rock chunks. Even the empty-looking Cassini division contains many particles.

A REASON FOR THE RINGS

No one is sure why Saturn has rings. The likeliest theory says that the rings are the remains of one or more shattered moons.

ORIGIN OF THE RINGS
Did Saturn gain its rings when a comet collided with one of its moons? The diagrams show a possible scenario.

Cosmic collision
A few hundred million years ago, a large comet or asteroid smashes into an icy moon orbiting Saturn.

Orbiting swarm
The impact shatters the moon into a cloud of icy particles. These then start to orbit Saturn.

Spreading out
Collisions among the particles grind them into smaller pieces. These spread out to encircle the planet.

Shepherd moons
The rings assume the shape we see today. They are gravitationally kept in place by small "shepherd" moons.

The Moons of Saturn

■ Saturn has at least 30 known satellites, ranging in size from mere rocks a few miles across to Titan, the second-largest moon in the Solar System.

THE MAIN MOONS

From Saturn's extended family, four moons stand out.

Titan At 3,200 miles (5,200 km) across, Titan is bigger than Mercury. It is unique among moons in having a thick atmosphere—mostly nitrogen with trace elements. Exposed to sunlight, the trace elements produce a smog which hides the moon's surface features.

Iapetus The moon Iapetus is known for its peculiar dark and bright hemispheres. The dark material is thought to be carbon-based, but astronomers do not know if it came from within Iapetus or was deposited from space. Iapetus is about 900 miles (1,400 km) in diameter.

Enceladus Enceladus has a diameter of 300 miles (500 km) and consists largely of water-ice. It appears to be the most geologically active of Saturn's moons. Its bright surface is a mixture of old, well-cratered areas and newer terrain that is grooved and fissured.

Mimas Mimas is fairly small, being 240 miles (390 km) across. It is pockmarked by impacts—a particularly large crater is the result of a collision that must have almost torn the moon apart.

TITAN'S HAZE This artist's impression shows Saturn viewed through the hazy atmosphere of its moon, Titan. No other moon has such a thick atmosphere.

Iapetus
Below: One side of this moon is very light; the other is very dark. No one knows why.

Titan
Right: Saturn's largest moon looks like a billiard ball—its surface is obscured by dense smog.

Mimas
The "bull's eye" on Mimas is the giant Herschel crater.

Enceladus
Much of this moon's surface may have been resurfaced by eruptions.

Uranus

■ Uranus is a blue-green world nearly four times the size of Earth. It was poorly understood until very recently.

BLUE-GREEN GAS GIANT

Most of Uranus is hydrogen and helium, like the Sun. And like the other gas-giant planets, it has no solid surface. In 1986, Voyager 2 saw a featureless, blue-green planet, but scientists using the Hubble Space Telescope are now seeing signs of storms in Uranus's upper atmosphere.

Blue wavelengths The blue-green color of Uranus comes from traces

TWO MOONS Two of Uranus's five major moons are Ariel (below left), 720 miles (1,160 km) in diameter, and Miranda (below right), 300 miles (485 km) across.

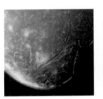

of methane in its atmosphere. The methane reflects the blue wavelengths of sunlight and absorbs the red.

MYSTERIOUS MOON

Many Uranian moons are small and asteroid-like, but five are quite large. One of them, Miranda, has a surface unlike any other moon in the Solar System. Images show a jumble of terrains, including cratered plains and grooved regions. No one knows how it came to look like this.

URANUS'S WORLD This infrared image of Uranus captures a wealth of detail, including the rings, some of the moons, and various cloud features.

FACT FILE

DISTANCE FROM SUN **19.2 AU**
SIDEREAL REVOLUTION PERIOD (ABOUT SUN) **84.0 years**
MASS (EARTH=1) **14.5**
RADIUS AT EQUATOR (EARTH=1) **4.0**
APPARENT SIZE **3–4 arcseconds**
SIDEREAL ROTATION PERIOD (AT EQUATOR) **17.2 hours, backward**
MOONS **21**

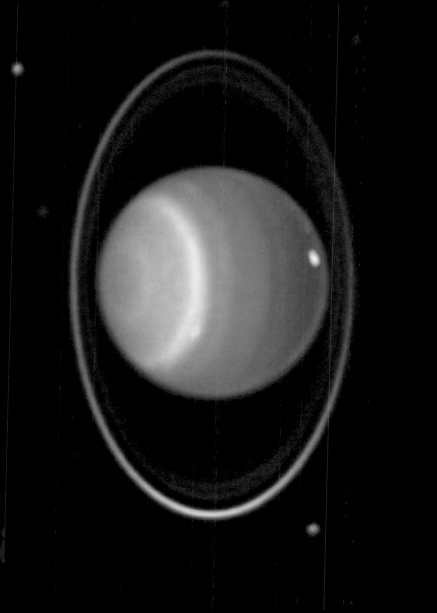

Neptune

■ Neptune is the smallest and most distant of the gas giants. It is also one of the most interesting, with a surprising amount of "weather" and a moon whose vents erupt nitrogen, not lava.

RAGING STORMS

Neptune's atmosphere is mostly hydrogen and helium gas with traces of methane. Temperatures in the upper atmosphere are so low that methane freezes.

Storms and clouds Neptune has raging storms, probably powered by a heat source deep within the planet. In 1989, Voyager 2 photographed a storm called the Great Dark Spot, which was nearly 6,000 miles (10,000 km) in length. It also saw a small white cloud, dubbed Scooter because of its rapid motion. Observations from Earth show that such storms come and go over the years.

RINGS AND MOONS

In 1984 astronomers found that Neptune had rings. Within this faint system are bright clumps, caused perhaps by moons that are yet to be discovered.

FACT FILE

DISTANCE FROM SUN **30.1 AU**
SIDEREAL REVOLUTION PERIOD (ABOUT SUN) **165 years**
MASS (EARTH=1) **17.1**
RADIUS AT EQUATOR (EARTH=1) **3.88**
APPARENT SIZE **2.5 arcseconds**
SIDEREAL ROTATION PERIOD (AT EQUATOR) **16.1 hours**
MOONS **8**

TRITON Triton is volcanically active, with geysers shooting nitrogen gas into the sky. With few craters, Triton's surface is covered in frosts of mostly nitrogen and methane.

We knew little about Neptune until the Voyager 2 flyby of 1989. The spacecraft took this shot of the planet's south pole (above), and discovered much about Neptune's moons, including Triton and Nereid (right).

Nereid

Triton

Two Neptunian moons were known before Voyager. Its flyby added six. By far the most fascinating is Triton.

Triton With a diameter of 1,700 miles (2,700 km), Triton is the largest Uranian moon and one of the largest in the Solar System—it is about two-thirds the size of our Moon and a little larger than Pluto. Its surface features range from the run-of-the-mill (impact craters) to the extraordinary (strange dimpled regions). In places, nitrogen gas shoots upward from vents for 5 miles (8 km) into the thin atmosphere of nitrogen vapor, before wafting downwind to form dark plumes on the ground.

Pluto

■ Pluto is a maverick. It is tiny and has a markedly eccentric orbit. Definitely not a gas giant, neither is it a terrestrial world. Where does Pluto fit in?

Pluto

OUT IN THE COLD

Pluto lies in the dark, cold hinterlands of planetary space. It takes 248 years for Pluto to orbit the Sun, 2.8 billion miles (4.5 billion km) away at closest approach. At the moment it is the farthest planet from the Sun, but because Pluto's orbit is so elongated it is sometimes nearer to the Sun than Neptune. Pluto's orbital path is also gently tilted from the plane in which the other planets orbit.

Rock and ice Being so far away, Pluto is the only planet that has not been visited by a spacecraft, so there is much that we do not know about it.

What we do know is that Pluto has a diameter of only 1,430 miles (2,300 km), making it the smallest planet—smaller even than our Moon. Pluto has a mass only one-fifth that of Earth and a rock-ice composition similar to Neptune's moon Triton.

Atmosphere Pluto's atmosphere may contain mostly nitrogen with some carbon monoxide and methane. It is very thin, however, and may exist as a gas only when Pluto is closest to the Sun. The planet's surface temperature varies between about −390° and −346°F (−230° and −210°C).

PLUTO'S ORIGINS

Pluto's unique features have led some astronomers to theorize that it came from the Kuiper Belt, a disk-shaped region lying beyond the zone of the planets. The bodies found there are icy planetesimals—comets without tails. Pluto may be the largest example of this group.

Charon

FACT FILE

DISTANCE FROM SUN **39.5 AU**
SIDEREAL REVOLUTION PERIOD
(ABOUT SUN) **248.0 years**
MASS (EARTH=1) **0.002**
RADIUS AT EQUATOR (EARTH=1)
0.18
APPARENT SIZE **0.04 arcseconds**
SIDEREAL ROTATION PERIOD (AT
EQUATOR) **6.39 days**
MOONS **1**

DOUBLE PLANET Charon, Pluto's moon,
is half the size of its planet, making the
two almost a double planet. Pluto's
surface may look like this (below): rocky
with a layer of nitrogen and methane ice.

Comets

■ When a comet is far from the Sun, it is a cold body, perhaps a few miles across, that looks like a large, dirty snowball. Astronomers believe that most reside in an enormous sphere of comets surrounding the Sun, known as the Oort Cloud, well beyond the orbit of the most distant planets.

FROM SNOWBALL TO COMET
Occasionally, one dirty snowball is jolted onto a path inward toward the brilliance of the Sun. As it closes in, the ice begins to boil away, and a head, or coma, of gas and dust develops. The material leaves the comet to form separate gas and dust tails streaming away from the Sun.
Periodic comets Sometimes a comet passes close to a planet, usually Jupiter, and the planet's gravity changes the comet's orbit. Repeated encounters may result in a new orbit that causes the comet to return over and over again to the inner Solar System. With its period of 76 years, Comet Halley is the best-known example of such a periodic comet.

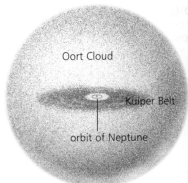

COMET ORIGINS Many comets come from the Oort Cloud, while others arrive from the Kuiper Belt, beyond Neptune.

COMET HALLEY Recorded sightings of Comet Halley go back more than [?]000 years. It last visited Earth's skies [in] 1986, when this photo was taken.

gas tail

dust tail

coma

nucleus

COMET STRUCTURE A typical comet has a tiny nucleus of ice and dust surrounded by a gaseous coma. Dust and gas stream from the coma in separate tails. The tails of a large comet might be tens of millions of miles long.

OTHER SOLAR SYSTEMS

Throughout history, the only planets that anyone knew about were Earth and its sisters orbiting the Sun. Then came the 1995 discovery of the first planet outside our Solar System. Scientists found a planet about half the mass of Jupiter orbiting a star called 51 Pegasi. Since then, more than 60 planets have been recognized, and there are many more tentative findings.

PLANET-HUNTING

The search is difficult because it is impossible to see these planets directly. Viewed from Earth, the planets appear close to their star, so they are lost in its glare. To find the planets, astronomers use indirect means. They try to detect little "wobbles" in the movement of a star caused by the gravity of an orbiting planet.

Orbiting close A look at the other planetary systems shows that they are built differently from the Sun's. Our Solar System has small-mass planets, such as Earth, orbiting near the Sun, while the largest planets, such as Jupiter,

orbit much farther out. But the other solar systems have large planets close to their star. Perhaps, as these solar systems were forming, something happened that moved the large planets inward.

NOTHING LIKE EARTH

So far, astronomers have found only massive planets with strong gravity orbiting fairly close to their star. These are planets about the size of Jupiter.

No one has seen what these planets look like, and nobody has found a planet anything like as small as Earth. But astronomers are hopeful that with improved equipment, Earth-size planets will be discovered too. Planets the size of Earth would be the most likely places to look for evidence of alien life.

ANOTHER SOLAR SYSTEM What would an alien planet system look like? All planets found around other stars are large bodies with masses roughly similar to Jupiter's. Astronomers believe that such planets would probably resemble Jupiter and have cloud belts and swirling storms.

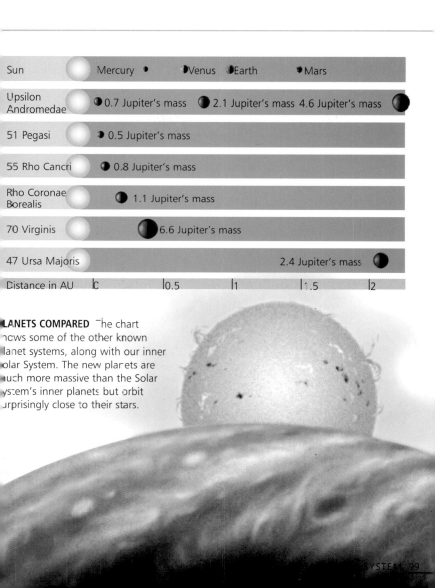

	Distance in AU				
Sun	Mercury •	•Venus	•Earth	• Mars	
Upsilon Andromedae	0.7 Jupiter's mass	2.1 Jupiter's mass	4.6 Jupiter's mass		
51 Pegasi	0.5 Jupiter's mass				
55 Rho Cancri	0.8 Jupiter's mass				
Rho Coronae Borealis	1.1 Jupiter's mass				
70 Virginis	6.6 Jupiter's mass				
47 Ursa Majoris				2.4 Jupiter's mass	
Distance in AU	0	0.5	1	1.5	2

PLANETS COMPARED The chart shows some of the other known planet systems, along with our inner Solar System. The new planets are much more massive than the Solar System's inner planets but orbit surprisingly close to their stars.

Reaching into the Universe

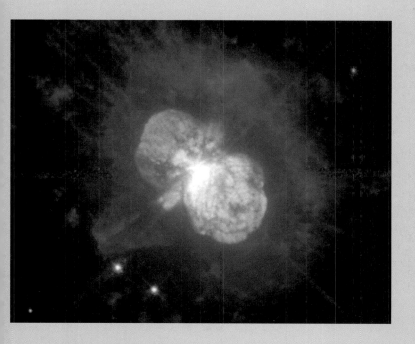

MODERN TECHNOLOGICAL developments have helped us paint a picture of a vast universe whose complexity would have astonished earlier observers. Bigger and better telescopes have been joined by new instruments—some of them in space—that can allow us to "see" invisible light.

LARGE TELESCOPES

The telescope has come a long way from its origins in the early 17th century. Galileo's first telescope was cruder than even a cheap pair of binoculars today, but it opened a new world.

EVER BIGGER Larger and larger telescopes followed William Herschel's 18th-century reflector (right). Its 48 inch (1.2 m) size compares with the 200 inch (5 m) built in the 1940s by George Hale (above).

TELESCOPE BUILDERS

The age of the giant telescope dawned in England in the 18th century. More powerful telescopes led to further discoveries.

First steps William Herschel discovered Uranus in 1781, and went on to build large telescopes that showed faint star clusters and nebulas. The Earl of Rosse built a 72 inch (1.8 m) reflector in 1845, and used it to discover the spiral structure of the galaxy M 51.

Hale's giants The early 1900s saw an explosive growth in the size of telescopes and in their light grasp

much of it due to George Hale,
a U.S. astronomer. Hale built four
"glass giants," culminating in the
200 inch (5 m) telescope on
Mount Palomar, California,
completed in 1948. It now bears
Hale's name to honor a lifetime's
contribution to astronomy.

EVEN BIGGER TELESCOPES
Cosmologists today use even
larger telescopes to map the
universe in ever greater detail.
The world's largest are the two
400 inch (10 m) Keck telescopes

LIGHT GATHERERS Astronomy's need for
more light—and bigger telescopes—is
inescapable because that is all the
universe sends us.

on Mauna Kea, Hawaii. Bigger
telescopes are being planned.

Computers now run observatories: They aim the telescopes,
operate the instruments, calibrate
the data, and keep the optics in
line. And computer-controlled
adaptive optics cancel out much
atmospheric distortion and
produce sharper images.

THE FULL SPECTRUM

The universe produces radiation over a wide spectrum. This radiation travels in waves of various lengths—from long radio waves, through infrared and optical waves, and on into shorter ultraviolet rays, X-rays, and gamma rays. Only a tiny part of the spectrum is visible.

VISIBLE LIGHT

The light we see is made up of a rainbow spectrum, each color representing a wavelength. A spectroscope breaks light from a star into a spectrum crossed by dark and bright spectral lines. By studying such lines, astronomers can determine many of the star's properties, including its composition and temperature.

The spectroscope also lets astronomers measure the speed of a star toward or away from us by noting a change in the wavelength of the lines. This is the Doppler shift (see diagram below).

LONGER WAVES

The low-frequency, low-energy part of the spectrum is the domain of infrared, millimeter, and radio radiation.

DOPPLER The principle of the Doppler shift allows astronomers to measure the direction of movement of stars and galaxies in relation to Earth.

1 No shift in the spectral lines: The galaxy is at rest.

2 A shift toward the red: The galaxy is moving away.

3 A shift toward the blue: The galaxy is approaching Earth.

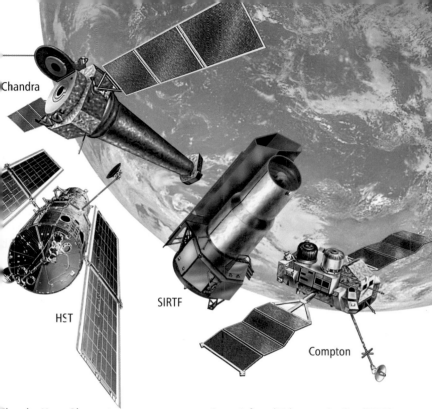

Chandra X-ray Observatory

The Chandra Observatory was launched in 1999, and is surveying the entire sky. It is looking for X-rays from objects such as distant exploding stars and merging clusters of galaxies.

Hubble Space Telescope (HST)

Instruments aboard the Hubble Space Telescope study the universe mainly in visual light. HST does, however, have the ability to capture images in near-infrared and ultraviolet wavelengths.

Space Infrared Telescope Facility (SIRTF)

This satellite observatory, due for launch in 2002, will study places where stars and planets are born. Such places give off radiation that falls mostly in the infrared region of the spectrum.

Compton Gamma-ray Observatory

This satellite mapped gamma-ray bursts until it was brought down from orbit in 2000 when systems began to fail. It will be replaced by the Gamma-ray Large Space Telescope, due for launch in 2005.

Infrared Infrared rays are the radiation we feel as heat. Most infrared sources radiate at temperatures between −430° and 1800° F (−260° and 1000° C). Infrared observations can pierce dense dust clouds to reveal young stars that are invisible at optical wavelengths, accentuate dusty disks around stars, and even pinpoint remote dusty galaxies.
Millimeter The millimeter window is ideal for observing giant molecular clouds in which stars are likely to form.

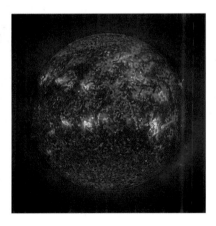

ULTRAVIOLET Observing the Sun at ultraviolet wavelengths highlights many details of the high-temperature solar corona, including sunspots and solar flares.

X-RAYS Purple and red colors indicate regions of high X-ray intensity in this Chandra X-ray Observatory image of NGC 3603, where stars are forming.

Radio Radio astronomy emerged as the first non-optical branch of astronomy. For more information, see pages 108–109.

SHORTER WAVES
At the high-frequency, high-energy end of the spectrum, ultraviolet, X-ray, and gamma-ray radiation unveil the physical and chemical properties of objects that are incredibly hot and energetic.
Ultraviolet light Observing at ultraviolet wavelengths allows astronomers to investigate objects such as gas around ordinary stars and hot stars whose evolution is running faster than the Sun's.
X-rays X-rays allow astronomers to study violent and extremely energetic objects and processes, such as supernova explosions.
Gamma rays Gamma rays provide a look at the oddest, most exotic objects in the universe, such as black holes, the swirling centers of active galaxies, and the Sun's hottest regions.

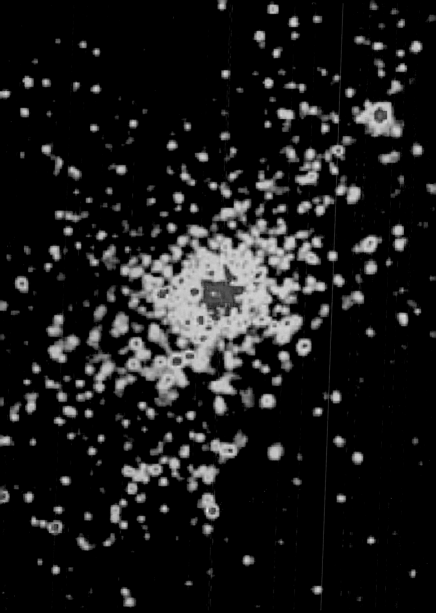

Radio Astronomy

■ Radio is a form of light with wavelengths longer than infrared. Radio astronomers tune in to these wavelengths to study some of the most important objects in the universe.

RADIO TELESCOPES
Most radio telescopes are large metal parabolic dishes, which focus the faint signals they receive from space. Big radio telescopes include the 250 foot (76 m) one at Jodrell Bank, England, and the 1,000 foot (300 m) antenna at Arecibo, Puerto Rico.

Interferometers To gain sharper views, astronomers link a number of radio telescopes together, creating, in effect, one giant-size telescope. Examples of these interferometers include the Very Large Array in New Mexico (27 antennas spread across many miles) and the Very Long Baseline Array, with 10 antennas from Hawaii to Puerto Rico.

DISCOVERIES
Radio astronomers discovered both pulsars, the spinning remnants of dying stars, and quasars, compact objects thought to be the active cores of very distant galaxies.

One of the most important discoveries came in 1965, when Arno Penzias and Robert Wilson of Bell Laboratories found an intriguing source of static. This was none other than the cosmic background radiation, the fading glow of the Big Bang.

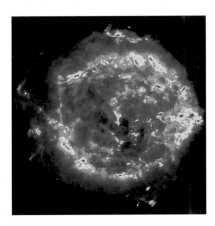

CASSIOPEIA A A radio image shows Cassiopeia A, an expanding shell of gas from a star that exploded as a supernova nearly 10,000 years ago.

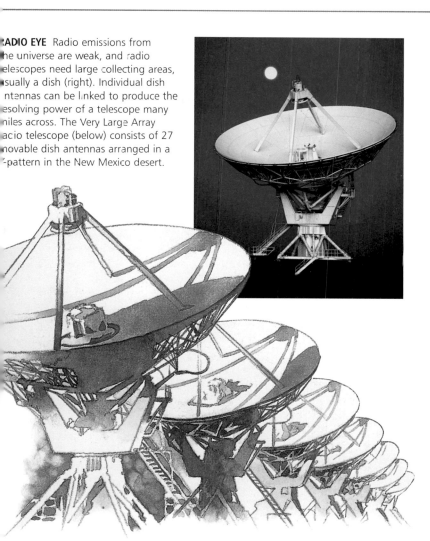

RADIO EYE Radio emissions from the universe are weak, and radio telescopes need large collecting areas, usually a dish (right). Individual dish antennas can be linked to produce the resolving power of a telescope many miles across. The Very Large Array radio telescope (below) consists of 27 movable dish antennas arranged in a Y-pattern in the New Mexico desert.

HUBBLE SPACE TELESCOPE

I n much the same way that
water distorts sunlight, Earth's
atmosphere distorts and filters the
cosmic radiation passing through
it. To overcome this problem,
astronomers some years ago
began lifting telescopes above
the atmosphere by placing them
in rockets and satellites. The
most famous of these orbiting
observatories is the Hubble
Space Telescope (HST).

ASTRONOMY FROM ORBIT

Launched in 1990, the Hubble
Space Telescope has provided
spectacular observations in
visual, near-infrared, and ultraviolet wavelengths, as well

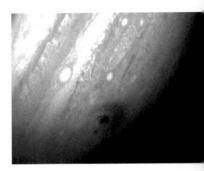

COMET IMPACT In July 1994, more than
20 fragments of Comet Shoemaker-Levy
struck Jupiter. HST's image of the planet
shows one of the smudgy impact areas.

as spectroscopic studies of stars,
the thin interstellar gas, and
galaxies. It consists of a 95 inch
(2.4 m) mirror and a suite of
sensitive scientific instruments.
All-seeing eyes The Wide Field/
Planetary Camera II, the most
often used of HST's instruments,
can detect objects as faint as 28th
magnitude (about a billion times
fainter than can be seen with the

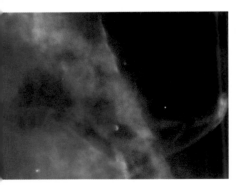

ORION'S PLUME HST zoomed in on part
of the Orion Nebula and snapped this
shot of a gas plume. Gas is condensing
into new stars within the nebula.

aked eye). The Faint Object Camera can also record 28th-magnitude objects, but it offers higher resolution and a wider choice of viewing angles.

Other instruments Two other instruments are the Near Infrared Camera and Multi-Object Spectrometer (NICMOS), and the Space Telescope Imaging

PRECIOUS CARGO The Hubble Space Telescope is prepared for deployment from the space shuttle in April 1990. HST orbits 380 miles (600 km) above Earth.

Spectrograph (STIS). NICMOS handles both imaging and spectroscopic observations of objects at near-infrared wavelengths. It aims to tells us much about the birth of stars in dense, dusty globules, the infrared emission produced by the active centers of distant galaxies, and the nature of a class of galaxies as bright as quasars at infrared wavelengths.

The STIS covers a broad range of wavelengths and can also block out the light of distant stars to search for black holes.

THE UNIVERSE IN CLOSE-UP

HST has looked at well over 14,000 objects and made more than 330,000 exposures. Out of these observations have come significant scientific insights. For example, astronomers now know more about how stars and stellar disks form, and that black holes exist at the centers of galaxies and quasars. HST has also increased our knowledge of the size and age of the universe, and detected galaxies that formed only a billion years after the Big Bang. Its high-resolution images of Mars, Jupiter, Saturn, and Neptune are surpassed only by space-probe photographs.

Detailed view The detail of HST's observations is so great that astronomers can now see,

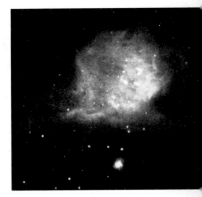

STELLAR VIEWS Stars are forming in this gas cloud in the galaxy NGC 6822 (above). A planetary nebula (NGC 6751) (right) in Aquila marks a stellar death.

for instance, great chunks of matter swirling around super-massive black holes at the centers of galaxies and quasars, as well as structural details in the spiral arms of nearby galaxies.

THE NEXT GENERATION

HST is now more than halfway through its planned lifetime. Scientists are planning its replacement, the Next Generation Space Telescope. This orbiting giant will have a mirror 26 feet (7.9 m) across and a light grasp more than 10 times that of HST.

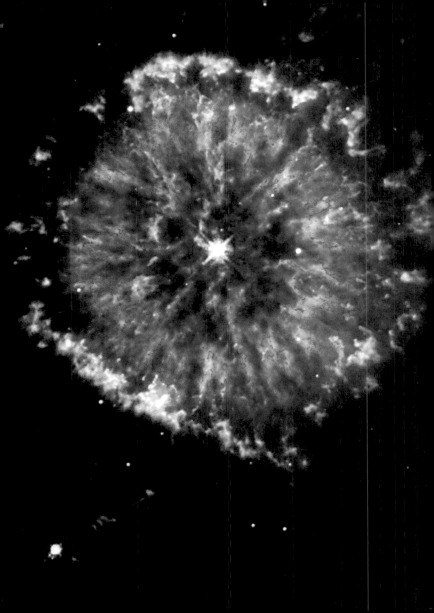

Space Exploration

THE TECHNOLOGY of the 20th century made the dream of exploring space a reality. Astronauts have walked on the Moon and lived in space stations, while robot probes have traveled throughout the Solar System. Future exploration may well answer the question: Are we alone?

HUMAN SPACEFLIGHT

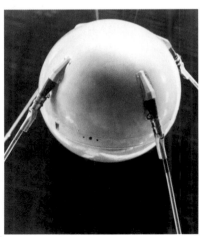

Dreams of space travel became reality in the late 1950s. As the Cold War between the United States and the Soviet Union deepened, the quest for superior spaceflight technology became another area of superpower rivalry. And a highly ambitious goal was set: putting a man on the Moon.

THE SPACE RACE

The Soviet Union took an early lead in what became known as the space race when they launched Sputnik 1 in October 1957. Sputnik 2 came a month later, carrying a live dog into space. And in April 1961 Yuri Gagarin became the first human in space when he made a single orbit of Earth aboard Vostok 1.

Three weeks after Gagarin's feat, Alan Shepard in a U.S. Mercury spacecraft traveled

STARTING WITH SPUTNIK The launch of Sputnik 1 (left) by the Soviet Union in 1957 gave Americans a profound shock. The U.S. fought back by establishing their own manned space program (top).

beyond Earth's atmosphere but did not go into orbit. John Glenn was the first American to reach orbit, in February 1962.

A WALK IN SPACE On June 3, 1965, Edward White in Gemini 4 became the first American to walk in space.

KENNEDY TARGETS THE MOON

In May 1961, President John F. Kennedy announced his country's goal of putting men on the Moon by the year 1970. To achieve this, the U.S. laid out a complex plan. After Mercury, a two-man Gemini craft would test techniques for joining spacecraft in orbit. It would also remain aloft for two weeks, the length of a lunar trip. Following Gemini, a three-man Apollo spacecraft, plus a gigantic booster rocket, would be developed to go to the Moon.

The Apollo Program

■ As 1966 ended, the Gemini program showed that U.S. spacecraft could change their orbits, which the Soviet craft could not. And Apollo's booster rocket, the Saturn V, was nearing flight tests while the Soviet lunar booster was still being planned.

THE COMING OF APOLLO

The Apollo program was almost stillborn when in 1967 a fire in an Apollo craft on the ground killed all three astronauts. After a pause,

TEST FLIGHT With Earth in the background, David Scott stands in the open hatch of Apollo 9, in 1969. This flight's mission was to test the lunar lander.

missions resumed in December 1968 with the launch of Apollo 8 by Saturn V rocket. It successfully circled the Moon for a day and returned to Earth.

Test runs The pace quickened. In March 1969, Apollo 9 tested its lunar lander in Earth orbit. This spidery craft was to take two astronauts from lunar orbit to the surface of the Moon. The two astronauts would then lift off in the upper part of the lander and rendezvous with the command module for return to Earth.

In May 1969, Apollo 10 performed a full dress-rehearsal: they traveled to the Moon, deployed the lunar lander, and took it to within 50,000 feet (about 15,000 m) of the surface.

SATURN V The success of the Apollo missions depended on the Saturn V rocket, which carried the crew, the lunar lander, and command module into space.

ONE GIANT LEAP

As Apollo reached its high point, the Soviet lunar program reached a new low. In early July 1969, the Soviets' lunar booster exploded on launch. Ten days later, they sent an unmanned spacecraft to the Moon, but the probe crashed.

July 16, 1969 saw Apollo 11 blast off with Neil Armstrong, Edwin Aldrin, and Michael Collins aboard. On the 20th, Armstrong and Aldrin set down on the Sea of Tranquillity. Armstrong said, "That's one small step for a man, one giant leap for mankind."

Aldrin and Armstrong spent two hours on the surface collecting rock samples and setting up experiments. Their safe return to Earth marked both the end of the space race and a great scientific triumph.

MEN ON THE MOON Safety worries sent Apollo 11 to the flat Sea of Tranquillity, mirrored in Edwin Aldrin's visor (right). Later Apollos explored more interesting sites, aided by a lunar buggy (above).

LATER LANDINGS
Five more landings followed. Apollo 12 stayed longer and collected more samples. Then came the drama of Apollo 13. Only heroic efforts and ingenious improvisation saved the crew's lives after a fuel tank exploded en route to the Moon. Apollo 14 restored U.S. confidence in lunar flight, and the last three missions visited sites showing complex geology. Apollo 17 brought an end to the program in December 1972, by which time a dozen humans had walked on the Moon.

The Space Shuttle

■ After the success of the Apollo missions, the U.S. space program shifted toward developing a reusable spacecraft. Dubbed the space shuttle, this craft could orbit Earth for two weeks at a time while carrying a cargo of more than 30 tons (30 tonnes).

TRIUMPH AND TRAGEDY

Columbia, the first shuttle, flew in April 1981. Three more shuttles were built and all of them flew routinely until January 1986, when Challenger exploded soon after launch, killing its crew. The fleet was grounded for two years.

WEIGHTLESS PEDALING A shuttle astronaut takes his turn on an exercise bike, part of the daily routine in the spacecraft's weightless world.

A JOURNEY BEGINS The shuttle Endeavour blasts off from Cape Canaveral. Each of its three main engines has enough thrust to power two and a half jumbo jets.

JUST A SPACE TRUCK

The shuttle is a "space truck." It delivers satellites into orbit and retrieves them for repair. Other payloads include scientific instrument packages for conducting experiments impossible on Earth.

Components The shuttle has three components: the orbiter, the external tank, and the solid rocket boosters. The orbiter is the delta-wing vehicle that carries a payload into orbit. The external tank fuels the shuttle engines during launch and burns up in the atmosphere. The two solid

ROOM TO SPARE The shuttle's payload bay is large enough to hold a bus. The huge mechanical arm at right is used to launch, repair, and service satellites.

rocket boosters that help propel the shuttle into space are ejected prior to orbit.

Accomplishments Among the shuttle's many accomplishments are launching probes to Venus and Jupiter, servicing the Hubble Space Telescope, and helping to build the International Space Station. It will continue to serve science for years to come.

Space Stations

■ A permanent human presence in orbit was a dream that began to come true in the early 1970s, when the Soviets launched Salyut, the first space station.

SALYUT TO MIR
Six successful Salyut missions were made between 1971 and 1986. In 1973 the U.S. launched Skylab, which made hundreds of experiments and observations before it was abandoned in 1974. **Mir** The most successful space station has been the Russian Mir, launched in 1986 and occupied almost continuously until 2000. One Mir crew member spent a record 437 days in space.

MULTINATIONAL EFFORT
The International Space Station is a project using components from many countries, including the U.S., Russia, Japan, and France.

Orbital assembly began in 1998, with new modules added over following years. Its first crew took up residence in November 2000, and a succession of crews should complete construction by 2005. The station can house up to seven people, who will conduct experiments in manufacturing, engineering, and technology. The station will also help scientists gain biomedical knowledge for manned expeditions into the Solar System.

VISIT TO MIR A space shuttle (foreground) is just about to dock with Mir. The Russian space station was visited a number of times by shuttles in a prelude to the construction of the International Space Station. Mir was brought down from orbit in March 2001.

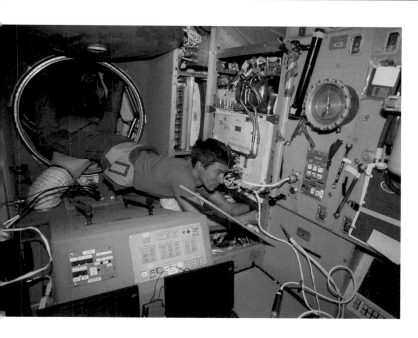

WEIGHTLESSNESS STUDY A Russian crew member of the International Space Station slides easily around the cabin in the weightless environment. The long-term effects of weightlessness—which include loss of muscle tone—will be closely studied aboard the station.

HOME IN SPACE When complete, the International Space Station will measure 356 by 290 feet (110 by 90 m), large enough for seven crew and equipment.

TO THE PLANETS

Human spaceflight is undeniably dramatic. Yet the most far-reaching discoveries about the planets have come from the data radioed back to Earth from unmanned spacecraft roaming the Solar System. Following a well-tested sequence—flybys first, then orbiters, then landers—scientists have used these robot probes to steadily increase our knowledge of Earth's neighbors.

TAKING AIM AT THE SUN

Missions that have focused on the Sun include the joint U.S. and European Solar and Heliospheric Observatory (SOHO), launched in 1995, which has studied the Sun's corona and activity from space. Ulysses (1990–), also U.S. and European, passes over the Sun's poles every five years to view it from directions we cannot see from Earth.

SCOUTING OUT MARS

Of all the planets, Mars has attracted the most attention from space agencies. When the U.S. Mariner 4 flew past in 1965, it saw a cratered landscape that resembled Earth's Moon. Later Mariners portrayed a somewhat different Mars—well cratered, but also having channels, volcanoes, and valleys.

Hoping to find life, NASA sent Viking 1 and 2 (1976), two pairs of orbiter and lander spacecraft. While the landers took soil samples and tested them for chemical signs of life, the orbiters took tens of thousands of photos, mapping the geology of Mars. The Viking landers found no evidence of life, past or present.

Pathfinder and Surveyor Following up on Viking, NASA's Mars Pathfinder lander and the Mars Global Surveyor orbiter arrived in 1997. Pathfinder carried a rover which drove around the lander, examining rocks. Global Surveyor's detailed images are rewriting the Mars geology book, as they point to a Mars that is

PATHFINDER Soon after touching down on Mars, the Pathfinder lander pictured a rocky red surface. The ramp (foreground) was used by the Sojourner rover,

SOHO
The Solar and Heliospheric Observatory (SOHO) takes ultraviolet photos of the Sun.

Sojourner rover
Rovers like Sojourner, carried by Mars Pathfinder, can explore wider areas of a planet's surface.

warmer, wetter, more volcanic, and geologically younger than scientists had thought.

VISITS TO HOT SPOTS

Spacecraft sent to the two planets closest to the Sun withstood scorching temperatures to radio back much of what we know about these intriguing worlds.
Mercury The planet Mercury has seen just one spacecraft, the U.S. Mariner 10, which made three flyby visits in 1974–5. Mercury was revealed as an airless Moon-like world of craters and lava plains, broiling under a Sun that makes the surface so hot (800°F, 430°C) that lead would flow like water.
Venus The first probe to reach another planet was the U.S. Mariner 2, which flew past Venus in 1962. While Mariner could not see through the clouds shrouding the planet, its instruments detected the hot surface and dense atmosphere. Venus soon became a favorite target for Soviet spacecraft. The Venera probes included landers that radioed back photos of a surface strewn with lava rocks and gravel.

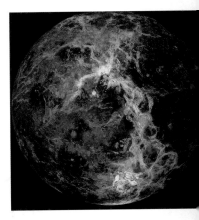

MAGELLAN'S VIEW Magellan used radar to pierce Venus's dense cloud cover and gain a detailed view of its surface.

To study the planet's features, radar was used to penetrate the cloud deck from orbit. Veneras 15 and 16 (both 1983) and the U.S. Pioneer Venus (1978–92) and Magellan (1990–94) missions mapped the volcanic landscape of Venus in great detail.

EXPLORING MICRO-WORLDS

Millions of asteroids, or minor planets, orbit between Mars and Jupiter. Only a few have been visited. Bound for Jupiter, NASA's Galileo probe flew past Gaspra (1991) and Ida (1993), discovering

TRIUMPHS Among the success stories in recent space exploration are Magellan (being launched from a space shuttle, below), which mapped Venus, and the Near Earth Asteroid Rendezvous (NEAR) spacecraft, which studied the asteroid Eros. NEAR sent this final image (left) just before hitting the asteroid's surface.

that the latter has a tiny moon, since named Dactyl.

The Near-Earth Asteroid Rendezvous (NEAR) craft visited Mathilde in 1997 on its way to orbiting Eros in 1999. In 2001, its mission complete, NEAR was deliberately crashed on Eros, becoming the first spacecraft to touch down on an asteroid.

JUPITER AND SATURN

Beyond the asteroid belt lie the gas giants. Flight times to these distant planets are measured in years, rather than in months as for the inner planets. Into this daunting realm only U.S. space-craft have ventured: Pioneers 10 and 11, Voyagers 1 and 2, Galileo, and Cassini.

OUT OF THIS WORLD

Voyager 1 is the most distant human-made object, at more than 7.4 billion miles (11.8 billion km) from Earth. Launched in 1977, it flew by Jupiter and Saturn in 1979–80, and is now nearing the edge of the Solar System. Voyager 1 should keep sending back data until at least 2020.

Jupiter Pioneer 10 made the first detailed portrait of Jupiter in a 1973 flyby, measuring the planet's powerful magnetic field and imaging its turbulent atmosphere. Pioneer 11 followed the year after, sharpening the picture.

The much more sophisticated Voyager probes flew past Jupiter in 1979, sending back fascinating images of the giant planet's atmosphere and its varied family of moons. Voyager 2 also discovered that Jupiter has a thin ring system.

In 1995, the Galileo spacecraft shot an entry probe into the Jovian atmosphere. As it fell, the probe's instruments sampled gases and measured wind speeds and temperatures. The main craft then spent several years surveying Jupiter and its moons.

Saturn Reaching Saturn in 1979, Pioneer 11 discovered new rings and drew the first close-up portrait of the Saturnian realm. Voyagers 1 and 2 arrived in 1980 and 1981. As at Jupiter, they detailed Saturn's features, finding new rings and moons and mapping the racing clouds of the planet's atmosphere.

RETURN TO JUPITER Jupiter, its colorful
cloud bands, the Great Red Spot, and the
moon Io, can be seen in this image taken
by Voyager 1 in 1979 (above). Just over
20 years later, another spacecraft—
Cassini, on its way to Saturn—captured Io
again, dwarfed by its mother planet (left).

VOYAGER 2 Voyager 2 was the first spacecraft to survey Uranus and Neptune. Much of what we know about these worlds comes from this single mission.

Neptune Voyager 2 completed its tour of the gas giants in 1989, when it reached Neptune. It found an active planet with giant storms in its atmosphere, and frosts and a thin atmosphere on its large moon, Triton. Its main mission completed, Voyager 2 moved out of the Solar System.

Cassini was launched in 1997 for a rendezvous with Saturn in 2004. It will orbit the ringed planet and also drop a probe onto Titan, Saturn's largest moon.

URANUS AND NEPTUNE

After their surveys of Saturn, Voyager 1 and both Pioneer probes headed out of the Solar System, while Voyager 2 took aim at Uranus and Neptune.

Uranus Arriving at Uranus in 1986, Voyager 2 found a feature-less planet shrouded in a blue haze of methane. In comparison to this seemingly bland world, the satellites of Uranus were of greater interest with their distinctive and complicated geologies.

TO THE EDGE

Only Pluto remains unvisited, more for budgetary reasons than technological ones. This tiny world poses many questions.

Comets that come relatively close to Earth are targets for study. The Stardust spacecraft is collecting particles from the tail of Comet Wild 2 for return to Earth. Among the missions currently on the drawing board are Contour (to fly past two or more comets between 2003 and 2006); Deep Impact (to fire a projectile at Comet Tempel 1 in 2005); and Rosetta (to arrive at Comet Wirtanen in 2011).

TAKING THE PLUNGE In 1995, the Galileo spacecraft's cone-shaped probe plunged into Jupiter's atmosphere. It radioed back information for about an hour before being crushed by the atmosphere.

GANYMEDE Jupiter's moon Ganymede fascinates scientists because of its diverse surface features. An image from Galileo shows impact craters on grooved terrain.

CASSINI The Cassini orbiter carries the Huygens probe, which will slowly descend through the smoggy atmosphere of Saturn's moon Titan in July 2004.

FUTURE MISSIONS

If just some of the plans being made by space agencies come to pass, the next century promises to be a new era of exploration.

MARS

There may be a veritable traffic jam of spacecraft in Martian skies in the next decade or so—all will be searching for signs of life. The European Space Agency's Mars Express orbiter and Beagle 2 lander will arrive in late 2004, just before a pair of NASA rovers. Later in the decade, NASA plans the first mission to return Martian rock and soil samples to Earth.

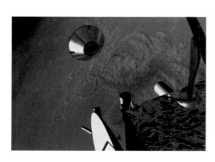

MARS EXPRESS Mars Express is scheduled for launch in June 2003 and will arrive in December 2003. This artists's impression shows it deploying the Beagle 2 lander.

BEAGLE 2 Once on the surface of Mars, the Beagle 2 probe will begin to collect samples from inside rocks (and under them) and test them for signs of life.

OTHER PLANNED MISSIONS

Mars is a hot-spot of future space exploration, but several other worlds will also receive visits. Here are just a few of the many planned missions.

Comets The Deep Impact mission plans to hit the nucleus of Comet Tempel 1 in mid-2005 with a projectile and study the results.

Europa Life may be present in an ocean beneath the icy skin of Jupiter's moon, Europa. To test this, NASA is planning a Europa Orbiter, to be launched in 2006, arriving four years later.

Asteroids MUSES-C, a Japanese mission, will visit asteroid 1998 SF36. The flightplan calls for MUSES-C to orbit the asteroid and pick up surface samples, returning them to Earth in 2007.

Mercury In 2008, a U.S. mission, Messenger, will fly to Mercury to complete Mariner 10's survey of the planet.

Some Future Missions

SPACECRAFT	COUNTRY	TARGET	ARRIVAL
NEAP	USA	asteroid Nereus	2002
Smart-1	Europe	the Moon	2002
Lunar-A	Japan	the Moon	2003
Contour	USA	Comet Encke; two others	2003+
Cassini/Huygens*	USA	Saturn's moon, Titan	2004
Stardust*	USA	Comet Wild 2	2004
MUSES-C	Japan	asteroid 1998 SF36	2005
Deep Impact	USA	Comet Tempel 1	2005
Messenger	USA	Mercury	2008
Europa Orbiter	USA	Jupiter's moon, Europa	2010

Indicates missions already in flight.

The Invasion of Mars

■ Establishing a colony on another world might be the next step for a species that seems eternally restless. Mars has always exerted a strong tug on the human imagination. How would a Martian colony establish itself? Here is one possible sequence of events, all based on the use of current technology.

SCOUTS COME FIRST

A scouting-out expedition would arrive well before the colonists. A scout rocket would contain both an Earth-return spacecraft and an atmosphere-processing factory. The factory would create methane and water using Martian air (carbon dioxide) and hydrogen brought from Earth. Methane and water could be converted into other necessities such as oxygen.

Leap-frog expeditions A crew of four to six astronauts would land two years after the rocket, and they would live on the surface in habitat modules. Launched toward Mars with them would be a second atmosphere-processing factory and Earth-return vehicle.

A LUNAR COLONY?

A human colony on the Moon is not yet deemed feasible, mainly on grounds of cost and on the Moon's lack of minerals needed on Earth. More likely is a scientific station, manned by crews in rotation.

The Moon would be an ideal site for astronomical observations. Radio telescopes could be sited on the lunar far side, shielded from Earth and its "chatter" of human broadcasting and telecommunications. Optical telescopes would also benefit from a lunar site. The lack of a filtering atmosphere means that they would enjoy the same, clear-sighted view of the universe as the orbiting Hubble Space Telescope.

This second vehicle would be set down elsewhere to begin making supplies for the next crew. The first astronauts would come back in the initial Earth-return spacecraft, leaving the scene set for the second expedition's arrival.

There would probably have to to be several such "leap-frog" expeditions before Mars was ready to receive its first colonists

COLONIES FOLLOW

A colony could develop from clusters of expeditions sent to one area. Left-behind habitat modules could be recycled into living quarters for a colony. The tougher problem is of building a Martian economy. This relies largely on what minerals are available on Mars and what is needed from Earth to convert

MARTIAN CITY This artist's conception of a human settlement on Mars may seem far-fetched, but the decades ahead may well see something like it come to pass.

those minerals into a usable form. In any case, given our present knowledge of Mars, little or no technological barrier stands in the way of creating a permanent human society on the Red Planet.

The Future of Space Travel

■ We are all familiar with TV images of rockets lifting off in clouds of smoke and sheets of flame. Launching rockets this way has been going on since the 1940s. Now, however, new technologies loom on the horizon. These promise much more efficient ways of moving people and payloads through space.

SUCCESSFUL TEST This artist's impression shows the 1999 flyby of the Braille asteroid by Deep Space 1. The mission successfully tested an ion drive.

ION DRIVES

NASA launched Deep Space 1 in 1998 to test an ion drive engine. This works by taking atoms, giving them an electrical charge, and accelerating them in an electric field. A stream of ionized atoms shoots out the back of the engine, producing a thrust which moves the spacecraft forward. An ion drive can run for months at full throttle.

Deep Space 1 and its ion drive worked well. Engineers are soon expecting to build drives that are much more powerful.

SOLAR SAILING

Another idea for moving through space is solar sailing—using the pressure of sunlight to blow a spacecraft from one planet to the next. Such a spacecraft would have a huge sail, perhaps $1/2$ mile (1 km) wide. Beyond Earth's atmosphere, the spacecraft's computers would unfurl the sail, made from highly reflective materials. Catching sunlight in this way gives the craft a gentle but neverending push.

GIVE NUKES A CHANCE

Another option is to develop a nuclear-powered rocket. The great advantage here is that the technology is well understood. And the benefits for spaceflight are clear-cut: a nuclear rocket offers a lot of power compared to a chemical one. A trip out to Mars, for example, might last as little as four months, half the time for a low-energy trip. The craft would also be reusable.

X-33 Developers of new spacecraft contend with an array of technological, financial, and political difficulties. The sleek X-33, for example, was to be a test bed for a new generation of reusable launch vehicles. After a run of problems, the project was cancelled by NASA in 2001 before the first test flight.

Nuclear rocket engines were in fact tested in the 1960s, but the program ended after public support for nuclear technology development evaporated.

SEARCHING FOR LIFE

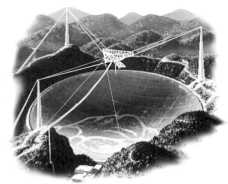

MESSAGE TO HERCULES In 1974, the Arecibo radio telescope sent a message to space about human life on Earth. It will take about 25,000 years to reach its target: the M 13 star cluster in Hercules.

MARTIAN METEORITE

Sometimes evidence of possible extraterrestrial life is found very close to hand. In 1996, scientists were studying a Martian meteorite that fell in Antarctica 13,000 years ago. They discovered what appeared to be fossilized bacteria that may have originated on Mars's ancient surface. Not everyone is convinced, however. Other scientists suggest that the tube-shaped structures may be the remains of compounds that entered the rock in Antarctica. The debate continues.

Are we alone? Of all the big questions astronomers grapple with, none is bigger than the inquiry into whether or not life exists elsewhere.

SOLAR SYSTEM SEARCH
Three places in the Solar System are deemed worthy of exploration for signs of life.
Mars The Red Planet has features that seem to have been formed by flowing water. If life did develop on Mars, these "wetlands" would be prime places to look.
Europa A similar watery world may exist on the Jovian moon Europa, beneath a layer of ice.
Titan Scientists want to probe Titan, Saturn's largest satellite. A variety of organic compounds—the building blocks of living things—have been detected in Titan's thick atmosphere.

SEARCHING THE COSMOS
The search for extraterrestrial life (SETI) began in the 1960s using radio telescopes to detect unusual signals from space. A more comprehensive search began in

the 1990s with Project Phoenix. Computers sift through natural signals from space to find signals that could only have been generated artificially. At any one time, tens of millions of channels can be monitored.

Earth calling In 1974, the Arecibo radio telescope sent a complex message written in binary code to M 13, the Hercules

A MARTIAN LAKE? A huge lake might once have filled the bluish area in this Viking image of Mars. Scientists are targeting such areas to search for life.

Cluster. The message included the chemical formulas for the molecular components of DNA. In effect, we were telling possible alien listeners what to expect if they dropped in for a visit.

OBSERVING
THE SKY

*Down-to-earth advice on
tools and techniques for every
backyard astronomer.*

Becoming a Stargazer

WE CAN DELIGHT in the wonders of the night sky just
by gazing upward. Binoculars open up new vistas, and
a telescope takes you into deep space. This chapter gives
a rundown on observing methods, and includes practical
tips on finding the best binoculars and telescopes.

How to Start Observing

To become a professional astronomer takes long training in physics and mathematics; to be a backyard skywatcher you simply go outdoors and look up.

PURSUING THE HOBBY

Astronomy can be pursued on a number of levels. Your curiosity determines how deep you will go.
Naked eye With no more than an astronomy guidebook, you can enjoy skywatching from any location. A beautiful grouping of the Moon and planets, or the progress of constellations throug the year—these are the simple pleasures of naked-eye astronom
Binoculars A pair of binoculars opens up exciting new vistas— from the Milky Way resolved int many thousands of stars, to the shuttling moons of Jupiter.
Telescopes If you do not yet ow a telescope, wait until you know your way around the sky. When you are ready to buy one, consul the guidelines on pp. 154–9. If you already have some equip- ment, the advice in this and oth chapters will help you make the most of it.

JOIN THE CLUB

Whether you become a serious observer or not, you will be able to share your interest in the sky with other amateur astronomers. Astronomy clubs are found in most cities and towns. Their meetings provide opportunities to meet fellow skywatchers, see a variety of telescopes in action, and get advice about what equipment to buy.

WHAT YOU CAN'T SEE

While bigger telescopes can show you more, be warned: no telescope will show nebulas and galaxies with the vivid colors depicted in long-exposure and computer-enhanced photographs. The human eye is simply not sensitive enough to see much color in faint deep-space objects, even through the lens of a large telescope.

Also, do not expect to see the flags or footprints on the Moon left behind by the Apollo astronauts. No telescope on Earth is powerful enough to show these.

EYEPIECE VIEWS The three illustrations below give an idea of what you will see in the eyepiece of a telescope.

The Moon
The Moon fills the eyepiece with its craters, hills, rays, and dark "seas."

Saturn
Saturn's rings are obvious in even a small telescope.

Trifid Nebula
Dark dust lanes run through this nebula's light-colored lobes of gas.

ON TARGET Telescopes range from simple ones such as this, suitable for beginners, to more complicated, computerized types.

Dark Skies and Good Seeing

■ Your observing site will have a greater effect on your skywatching than any piece of equipment.

CHOOSING A SITE

Unshielded street lighting paints a glaring skyglow above every city and town. A dark site, away from urban lights, is ideal.

Urban skies Fortunately for city-bound stargazers, the Moon and planets remain unaffected by light pollution and provide wonderful targets. Even so, try to find a site that is shadowed from street and yard lights.

Head for the hills For the best views of the Milky Way, its star clusters and nebulas, and of faint galaxies beyond, plan to travel away from city lights, perhaps to a park or conservation area. Haze-free and moonless nights (around New Moon) are best—moonlight can wash out faint objects as effectively as can streetlight.

CLEAR SKIES, GOOD SEEING

If the sky is clear, the Milky Way may be seen even from suburban locations, and you are more likely to find some faint objects such as nebulas. Moisture degrades the transparency of the sky.

However, memorable nights of viewing come with many kinds of weather. For the Moon and planets, humid nights with some haze or thin fog can actually bring the sharpest views. Even though little else is visible in the sky, planet disks appear absolutely steady, revealing astonishing detail. This is known as good seeing—it happens when the layers of Earth's atmosphere are calm and stable, and not mixed up by winds at different altitudes. Under turbulent conditions, the poor seeing turns the disks of planets into boiling blobs with, at best, some fleeting moments of sharp views.

Mix and match At most sites, nights of good seeing are often the least transparent, and vice versa. Backyard astronomers soon learn to adapt their observing priorities to the conditions of the night, perhaps pursuing a galaxy hunt on a clear night, and planet studies on nights of good seeing.

...GHT POLLUTION Mode-n
-ilization, with its glow of
-tific al lighting, puts barriers
.. the way of appreciating the
-y. The scale of the problem
-n be clearly seen in this
-tellite image of the night
-hts of the eastern Uni-ed
-ates (above). In the most
-nse y populated regions,
-w lccations can be said to ke
-ly dark. The true beauty of
-e night sky only becomes
-parent at a dark site (right).

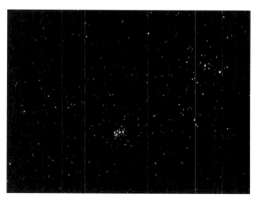

BINOCULARS

craters. A bright comet is often best seen with binoculars, as are solar and lunar eclipses, and close gatherings of the Moon and planets at twilight.

FINDING THE RIGHT TYPE

Binoculars are identified by a figure such as 7 x 50. The first number refers to the magnification, the second to aperture.

Magnification In the example above, the magnification is 7 power (7x). Note that binoculars with high magnification (16 or 20 power) have a narrow field of view, making it difficult to find objects. They are also hard to hold steady.

T he low power, wide field of view, and upright images of binoculars make it a snap to find celestial targets. Spend a year exploring the sky with binoculars and you will be better prepared to find objects with a telescope later.

WHAT YOU WILL SEE

From a dark-sky site, binoculars let you see bright star clusters, many nebulas, several galaxies (including the beautiful Andromeda Galaxy), and many star-packed regions along the Milky Way. Binoculars can also reveal the moons of Jupiter and the largest Moon

THE RIGHT TYPE The most popular binoculars for astronomy have front lenses 50 mm across (top). Make sure they are comfortable to hold (below).

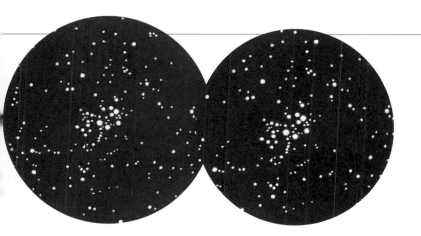

Aperture The second number gives the aperture, in millimeters, of each of the twin front lenses. Compared to 35 mm or 42 mm models, binoculars with 50 mm lenses gather more light and provide brighter images, important for tracking down faint objects in the night sky. Models with larger lenses are heavy and hard to hold. **Recommended models** The best choice is either a 7 x 50 or a 10 x 50 pair. Each offers a good balance of power, image brightness, and light weight. Avoid fixed-focus or zoom binoculars, which provide inadequate image quality for viewing sharp, pinpoint stars.

DIFFERING VIEWS These two views of the Pleiades show the relationship between field of view and magnification. Most 7x binoculars show about 7 degrees of sky (above left). Higher-power 10x models magnify the image more, but usually show only 5 degrees of sky (above right).

FIELD OF VIEW

Most 7-power binoculars provide a field of view of 7 degrees, enough to take in the Pointer Stars of the Big Dipper's bowl, or all of Crux, the Southern Cross. Higher power 10x models usually have a smaller field of view, perhaps 5 degrees, still sufficient to provide impressive sky views.

Some binoculars offer wide-angle eyepieces that provide a

larger field of view than normal, perhaps 7 to 8 degrees on a 10x model. However, image quality often suffers in such models.

HOLD STEADY!

Holding binoculars as steady as possible improves the view, allowing you to see the stars as steady points of light, rather than as a bunch of flitting fireflies. One way to ensure steadiness is to lie

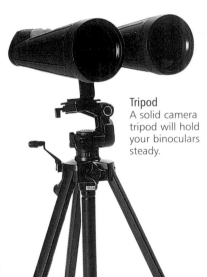

Tripod
A solid camera tripod will hold your binoculars steady.

on a reclining deck chair and prop your arms and binoculars on the chair's arms. You could also simply lean against a wall or fence. As an alternative, many binoculars can be attached to a

camera tripod, but make sure the tripod is a sturdy one.

Cantilever arms Tripods help with steadiness, but are not much good for scanning directly overhead. For comfortable views of objects high in the sky, consider purchasing a stand with cantilever arms. The arms swing the binoculars away from the tripod allowing you to stand beneath the binoculars.

TELESCOPES

A telescope can show you craters on the Moon as small as 1/2 mile (about 1 km) across, Jupiter's clouds, Saturn's rings, as well as star clusters, nebulas, and galaxies that are too small and faint to show up in binoculars.

TELESCOPE SAVVY

A telescope's most important specification is its aperture—the diameter of its main lens or mirror. The larger the aperture, the brighter and sharper the images. A 6 inch (150 mm) telescope will provide images twice as sharp and four times brighter than will a 3 inch (75 mm) telescope.

Magnification The least important specification is magnification. By changing eyepieces, any telescope can be made to magnify any amount.

TYPES OF TELESCOPE

There are three basic kinds of telescope: refractor, reflector, and catadioptric.

Refractors Refractors with an aperture of more than 4 inches (100 mm) are large, heavy, and expensive, so this type of telescope is most popular in the entry-level 2.4 inch (60 mm) to 3.5 inch (90 mm) aperture sizes. These models produce sharp images of the Moon and planets, but their small apertures limit their use for hunting faint, deep-sky targets.

Reflector Reflectors lack the expensive lenses of refractors, allowing a 4.5 inch (110 mm) or 6 inch (150 mm) reflector to sell for the same price as a smaller refractor. Because of their greater aperture, reflectors are a good choice for viewing nebulas and galaxies from a dark-sky site.

Catadioptric Catadioptric telescopes offer fine optics and generous light-gathering power in compact tubes much shorter than most refractors and reflectors.

MOUNTS

Make sure that the telescope is well mounted. A wobbly stand will produce images that never stop bouncing, making for frustrating viewing.

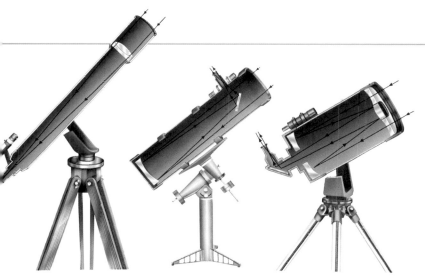

Refractor
This uses a lens at one end to gather light, which is focused to an eyepiece at the other end.

Reflector
A mirror collects light, which is reflected back up the tube to the eyepiece.

Catadioptric
Light travels back and forth between a mirror and a lens before entering the eyepiece.

Types Mounts come in several varieties. Altazimuth types are easy to set up but cannot follow the stars with one simple motion. Dobsonian mounts, a variation on the altazimuth design, need only gentle nudges to keep objects centered. The more sophisticated equatorial mounts permit motorized tracking of the stars. Fork mounts are standard on many catadioptric telescopes.

WHAT TO BUY

A 6 inch (150 mm) or 8 inch (200 mm) reflector teamed with a Dobsonian mount is one of the best buys for the serious stargazer. You get generous aperture on a rock-steady mount that is simple to set up.

At a similar cost, 4.5 inch (110 mm) reflectors on equatorial mounts have long been popular as starter telescopes. The mounts are

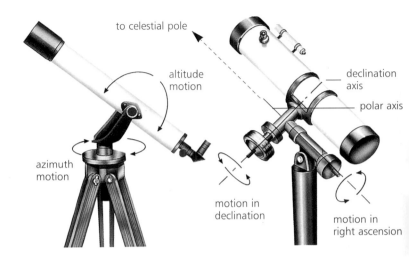

to celestial pole

altitude motion

azimuth motion

declination axis

polar axis

motion in declination

motion in right ascension

KNOW YOUR MOUNT The altazimuth mount (left) is widely used on small refractors and is also found on many reflectors. The equatorial mount (right) is used on many small reflectors.

more complex to set up (they must be aligned to the celestial pole to track properly), but do make it easier to follow objects. Look for models with "heavy-duty" or premium-grade mounts.

Some people prefer to select a refractor because it needs less maintenance than a reflector. A 2.7 inch (70 mm) to 3.5 inch (90 mm) refractor, provided it

has a solid altazimuth or equatorial mount, offers sharp optics in a rugged package that rarely needs upkeep. Avoid 2.4 inch (60 mm) refractors, as few models in this size offer good-quality mounts and fittings.

If portability is very important to you, then a catadioptric makes a good, though more costly, choice. Popular models include 3.5 inch (90 mm) and 5 inch (127 mm) Maksutov-Cassegrain and Schmidt-Cassegrain telescopes. The 8 inch (200 mm) Schmidt-Cassegrain models have long been a top-selling

Good choice
A good choice is this reflector on an equatorial mount.

Computer-controlled
At the touch of a button, a computer-controlled telescope locates even the most obscure of galaxies.

Solid mount
The solid, high-quality mount offers rock-steady viewing.

choice of amateur astronomers looking for sizable aperture in a compact telescope.

EYEPIECES

You change magnifications by changing eyepieces. A basic set of three eyepieces might include a 25 mm (for low-power views of deep-sky objects), an 18 to 12 mm (for medium-power views of the Moon and star clusters), and a 10 to 7 mm (for high-power views of planets and double stars). Invest in quality eyepieces, such as Kellners and Modified Achromats. A step up takes you to Plössl eyepieces.

An alternative to high-power eyepieces is to add a Barlow lens. One of these will double or triple the power of any eyepiece inserted into it, while retaining the comfortable eye relief inherent in all low-power eyepieces.

TELESCOPE TARGET Through a medium to large telescope, the Ring Nebula resembles a ghostly smoke ring. A larger instrument can pick out the central star.

Accessories
Accessories include eyepieces (back), colored filters, a Barlow lens (lower left) and a camera adaptor (right), which couples a camera to a telescope.

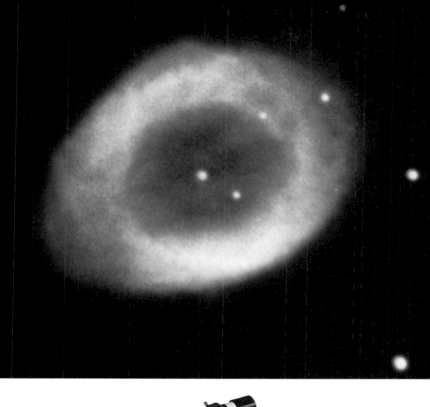

EYEPIECES

A telescope's main lens or mirror gathers the incoming light and focuses it into an image, but it is the eyepiece that magnifies the image. To change magnifications, you need to change eyepieces.

SOLAR SAFETY

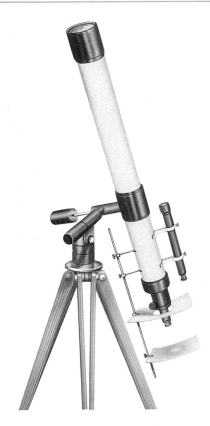

The Sun is a source of endless fascination, but looking directly at its disk can cause blindness. A few precautions will ensure safe viewing.

TWO METHODS

In the past, many entry-level telescopes came with "sun filters" that screwed into an eyepiece. These are very dangerous and should never be used. Instead, try one of these two methods.

Solar filter The easiest way to view the Sun safely is to use a solar filter that fits over the telescope's aperture. Cover the finderscope to prevent scorching someone by accident.

Sun screen Alternatively, use a Sun projection screen. First cap the finderscope, then position the telescope by moving it until its shadow is at a minimum. The Sun should be shining through the telescope and out through the eyepiece to the ground. Then project the Sun's image onto a piece of paper, shielding this screen from direct sunlight. Focus until the image is sharp.

SAFE VIEWING This telescope is equipped with a Sun projection screen. The Sun's image is focused onto the lower screen, while the upper screen shields the image from other sunlight. An advantage of using this method is that it enables a group of people to view the Sun safely.

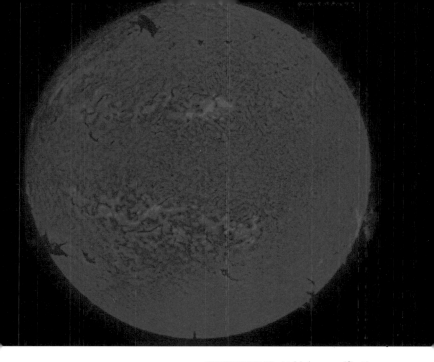

SUN'S SURFACE At high magnification, the Sun's surface, the photosphere, has a granular appearance, like oatmeal.

ola⁻ filter

he arge solar filter dwarfs ⁻he
olo⁻ed filters, which a⁻e used
or viewing planets.

ASTROPHOTOGRAPHY

Hardly anyone who has a camera and an interest in photography can resist taking photos of the sky. And capturing the stars does not necessarily mean investing in expensive cameras and films or using difficult techniques.

CAMERAS AND FILMS

For casual sky-shooting, almost any design of camera will do, but it must have a B (for Bulb) setting for the shutter. This is used with a locking cable release to hold the shutter open—most night-sky photos require long exposures.

AIMING AT THE SKY
Cameras with simple mechanical shutters work more reliably than electronic models whose batteries can die during long exposures.

Night shots need the extra light-gathering ability of fast film to bring down exposure times. Use at least an ISO 400 film.

TECHNIQUES

There are two main techniques. One uses a camera on a tripod; a more sophisticated method uses an equatorial mount to track the stars during an exposure.

Camera-on-tripod Load a camera with fast ISO 400 to 800 film, place it on a tripod, and focus the lens to infinity. Set the lens aperture to f/2.8, then frame a constellation. Use a cable release to hold the shutter open for 20 to 40 seconds and you will record as many stars as your eyes can see.

Tracking shots Recording more stars or faint nebulosity requires a dark site and bolting the camera to a motor-driven equatorial mount that can accurately track the stars during a 5- to 15-minute exposure. Your telescope may already have an attachment bolt, or you can buy a bracket. The results of this "piggyback" technique can be spectacular.

STAR TRAILS From a dark site, locking the shutter open for 10 to 60 minutes creates star trail portraits as Earth's motion causes the stars to make streaks on the film.

PIN-POINT SKY Using the piggyback method, stars will record as points, even in 10- to 30-minute exposures. But the telescope must be aligned as closely as possible to the celestial pole.

Understanding the Sky

LEARN ABOUT THE SKY and how it moves; about the celestial sphere and astronomical coordinates; about star magnitudes; about finding your way around the heavens and using the best observing techniques.

MEASURING THE SKY

Though we know Earth does the moving, it can still be convenient to picture the sky as the ancients did, as a crystalline sphere turning above our heads with the sky's contents "pasted" to its interior surface.

THE CELESTIAL SPHERE

The celestial sphere has an equator and two poles, like Earth. Coordinates similar to latitude and longitude allow astronomers to locate the positions of heavenly bodies accurately.

Declination As latitude measures distance north or south of Earth's equator, declination measures angular distance from the celestial equator. It runs from 0 degrees at the equator to 90 degrees (north and south) at the poles. One degree (°) of declination contains 60 arcminutes ('), and each arcminute contains 60 arcseconds (").

Right ascension The celestial equivalent of longitude is right ascension, or RA. It measures how far east or west a star is. The sky is divided up into 24 hours (h) of RA, each hour containing 60 minutes (m) of time. The equivalent of Earth's 0-degree Greenwich meridian is the point on the celestial equator where the Sun stands each year on the date of the March equinox. The line of right ascension that intersects the celestial equator at this point is defined as 0 hours RA.

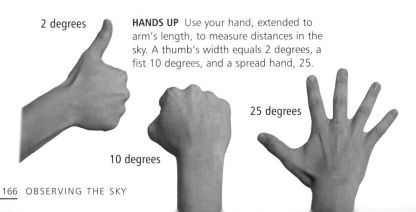

HANDS UP Use your hand, extended to arm's length, to measure distances in the sky. A thumb's width equals 2 degrees, a fist 10 degrees, and a spread hand, 25.

2 degrees

10 degrees

25 degrees

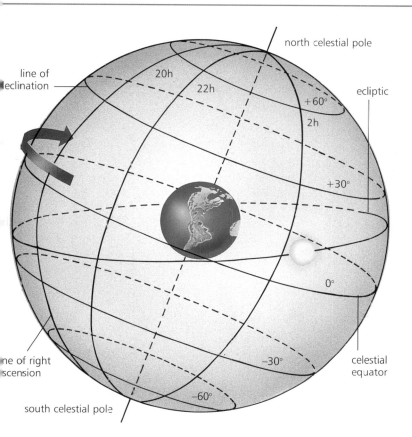

north celestial pole

line of declination

ecliptic

20h

22h

+60°

2h

+30°

0°

line of right ascension

−30°

celestial equator

south celestial pole

−60°

CELESTIAL SPHERE The positions of stars are described by coordinates on an imaginary sphere. The ecliptic is the Sun's apparent path across the background of the celestial sphere. The red arrow indicates the sphere's apparent daily movement westward.

The Motion of the Stars

■ At first glance, the stars seem to be fixed in place in the night sky, immovable. But a little observation over just a few hours will show that this is not so.

THE CHANGING SCENE

As the world turns, stars rise above the eastern horizon and set below the western horizon. On a December evening, for example, you might see Orion rising in the east. But by midnight Orion stands high in the sky. By dawn, it is setting in the west.

As Earth revolves around the Sun, the night side of the planet looks out toward a changing array of constellations. In June, we look toward Sagittarius. Orion lies in the opposite direction, near the Sun in the daytime sky. Six months later in December, the night side of Earth looks toward Orion while the Sun appears to be in Sagittarius.

CHANGES WITH LATITUDE

As Earth spins, the sky appears to rotate around its two celestial poles. Just how the stars move in relation to your horizon depends on where on Earth you live (see diagrams, below).

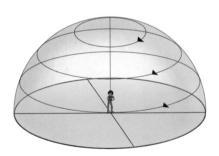

At the North Pole (90°N)
The north celestial pole is directly overhead. Only stars in the northern half of the celestial sphere are visible.

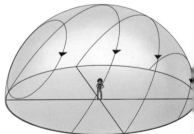

At northern middle latitudes
Part of the sky is always invisible, and stars turn in circles centered on the celestial pole, lying due north.

SLOW MOVEMENT Earth's rotation makes the sky appear to move from east to west. The illustration shows Orion as seen looking south.

GRADUAL CHANGE Two weeks later, at the same place and same time of night, you can see how Orion has moved slightly to the west.

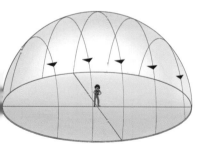

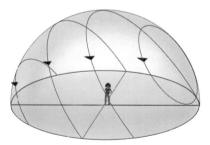

At the Equator (0°)
The celestial equator is directly overhead. The stars rise straight up and sink straight down again below the horizon.

At southern middle latitudes
Part of the sky is always invisible, and stars turn in circles centered on the celestial pole, lying due south.

STAR BRIGHTNESS

The first thing you notice on looking skyward is that stars vary in brightness. They do so for two reasons: some are closer to us than others, and some really are brighter than others.

TWO MAGNITUDES

The concepts of absolute and apparent magnitude allow us to describe star brightness precisely.

Apparent magnitude This indicates how bright a star appears to the naked eye—the lower the magnitude, the brighter the star. For example, the brightest star in the sky, Sirius, shines at about −1.5 magnitude, while the faintest stars visible to the naked eye in a dark sky are about magnitude 6. A difference of five magnitudes equals a difference of a hundred times in brightness.

Absolute magnitude The nearer the star is to Earth, the brighter it will appear, and since stars also shine with different luminosities, apparent magnitude does not measure the true brightness of a star. In order to describe intrinsic brightness, astronomers define absolute magnitude as the apparent magnitude a star would have if it were 10 parsecs or 32.6 light-years from us.

For example, the Sun's apparent magnitude is −26.8, but its absolute magnitude is 4.8. Sirius is intrinsically much brighter than the Sun: it has an absolute magnitude of 1.4.

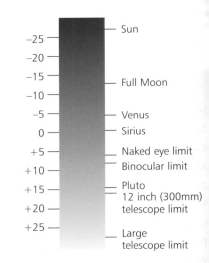

APPARENT MAGNITUDE SCALE The apparent magnitude scale describes how objects *appear* in our sky. It does not describe the objects' true brightness.

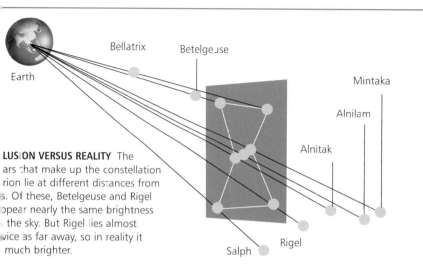

ILLUSION VERSUS REALITY The stars that make up the constellation Orion lie at different distances from us. Of these, Betelgeuse and Rigel appear nearly the same brightness in the sky. But Rigel lies almost twice as far away, so in reality it is much brighter.

The 10 Brightest Stars

COMMON NAME	CONSTELLATION NAME	APPARENT MAGNITUDE
Sirius	α Canis Majoris	−1.46
Canopus	α Carinae	0.72
Alpha Centauri	α Centauri	0.01
Arcturus	α Boötis	0.04
Vega	α Lyrae	0.03
Capella	α Aurigae	0.08
Rigel	β Orionis	0.12
Procyon	α Canis Minoris	0.8
Achernar	α Eridani	0.46
Hadar	β Centauri	0.66

NAMES IN THE SKY

The night sky is divided into constellations and known stars are given names or numbers.

CONSTELLATIONS

Constellations are just join-the-dot patterns in the sky invented by skywatchers through the ages. There are 88 of them, their borders defined by astronomers in 1930. The boundaries may be modern, but many constellations are ancient. Taurus dates from the 3rd millennium BC. Others, such as

Tucana

WHAT'S IN A NAME? Many stars and deep-sky objects have more than one name. The Orion Nebula is also known as the Great Nebula, M 42, and NGC 1976.

Tucana, are inventions of 17th- and 18th-century astronomers.

HOW STARS ARE NAMED

The names we use for stars come mainly from early Greek and Arab astronomers. But several other naming systems are also used.
Bayer letters Greek letters are applied to stars in a constellation so that the brightest star is generally called alpha (α), the

The Greek Alphabet

SYMBOL	NAME	SYMBOL	NAME	SYMBOL	NAME
α	alpha	ι	iota	ρ	rho
β	beta	κ	kappa	σ	sigma
γ	gamma	λ	lambda	τ	tau
δ	delta	μ	mu	ν	upsilon
ε	epsilon	ϑ	nu	φ	phi
ζ	zeta	ξ	xi	χ	chi
η	eta	ο	omicron	ψ	psi
θ	theta	π	pi	ω	omega

ext brightest beta (β), and so on.
Betelgeuse in Orion is therefore
also called Alpha (α) Orionis,
Orionis being the genitive form
of the constellation name.

Flamsteed numbers Stars are
numbered west to east across a
constellation, so Betelgeuse is
also 58 Orionis.

M objects These are star clusters,
nebulas, and galaxies in the
Messier list, compiled by the
18th-century French comet-
hunter Charles Messier.

NGC objects These are star
clusters, nebulas, and galaxies

Taurus

listed in the New General
Catalogue of J. E. L. Dreyer,
published in the late 1880s. The
NGC, along with its two index
catalogs (IC), lists more than
13,000 objects.

FINDING YOUR WAY AROUND

F inding your way around the sky can seem daunting for a novice, but it is really no harder than reading a road map. The first step when observing is to get oriented. To make sense of star charts you need to know which way is north and south. And for telescopes with equatorial mounts to track properly they must be aligned to the celestial pole in your hemisphere.

FINDING NORTH
Northern Hemisphere dwellers are fortunate in that the sky provides a bright star close to the north celestial pole. This is Polaris, the pole star.

There is also a convenient way to find it. Just locate the Big Dipper (in Ursa Major), mentally draw a line joining the two stars at the end of the bowl, extend it five times, and you have arrived at Polaris.

FINDING SOUTH
For Southern Hemisphere sky-watchers, finding south is not quite so easy. But although the

south celestial pole is not marked by a bright star, there are several ways to find the pole.

The easiest method involves first locating Crux, the Southern Cross. Its long axis points to the south celestial pole, a blank area marked by the dim star Sigma Octantis. The pole lies halfway between Crux and the bright star Achernar and forms a right-angle triangle with brighter Canopus. A line perpendicular to a line joining Alpha and Beta Centauri also points to the pole.

BRIGHT CONSTELLATIONS
In finding your way around the sky, it rapidly becomes apparent that a few constellations are bright and easy to find. These can be used as jumping-off points on your journeys from star to star.

For example, Orion is prominent in the sky from most locations in the first few months of the year. More or less opposite Orion in the sky is Scorpius, which is prominent in the northern summer (southern winter) sky. Another bright

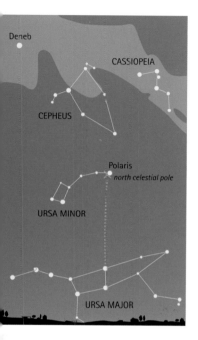

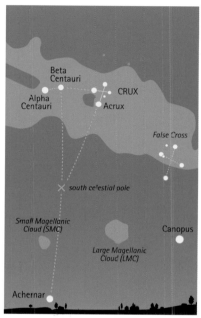

FINDING NORTH Draw a line through the two stars at the end of the Big Dipper's bowl, then extend it to Polaris.

FINDING SOUTH With no bright "south star" to mark due south, use Crux to point to the sky's southern pole.

constellation, familiar to most southern observers, is Crux, the Southern Cross.

Other landmarks Some prominent groupings of stars are not complete constellations. These

are called asterisms. Examples include the Big Dipper in Ursa Major, the W shape of Cassiopeia, the "teapot" of Sagittarius, the Great Square of Pegasus, and the sickle of Leo.

ECLIPSES

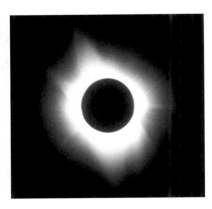

SOLAR ECLIPSE The Moon eclipses the Sun by passing in front of it and casting its shadow on Earth (below). During a total solar eclipse (above), the Moon completely covers the Sun's disk.

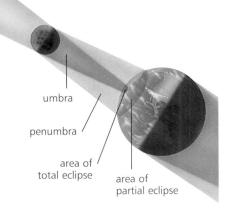

umbra

penumbra

area of total eclipse

area of partial eclipse

I n one of the great coincidences in nature, the Sun, 400 times the size of the Moon, also lies 400 times farther away than the Moon. As a result, both objects appear to be the same size in our sky, allowing the Moon to neatly cover the Sun in a total eclipse.

ECLIPSES OF THE SUN

Each month at New Moon phase, the Moon passes between us and the Sun. The tilt of its orbit usually takes the Moon above or below the Sun's disk, so it misses eclipsing the Sun. But about twice a year the Moon's orbit intercepts the Sun just as the Moon reaches New phase. The three worlds fall into alignment with the Moon in the middle, and we get an eclipse of the Sun.

The eclipse might be partial, when the Moon covers only a portion of the Sun. Or it could be annular: if the Moon is at the

DIAMOND RING As totality approaches during a solar eclipse, a last burst of sunlight shines through a valley on the Moon, creating a "diamond ring" effect

most distant point in its orbit, its disk will not be large enough to completely cover the Sun. We see the Sun reduced to a ring of light.

The most spectacular eclipse occurs when the Moon totally covers the Sun's disk. The Moon's dark umbral shadow sweeps across Earth along a path several thousand miles long. Within this track people see a total eclipse.

ECLIPSES OF THE MOON

At least twice a year the Sun, Earth, and Moon align with Earth in the middle. The Moon then passes through our planet's shadow, creating a lunar eclipse.

During a penumbral eclipse, the Moon passes through just the

A RUDDY MOON The color of a totally eclipsed Moon comes from sunlight refracted through Earth's atmosphere, staining the gray lunar surface with the hues of sunrise and sunset.

outer portion of Earth's shadow, and the Moon darkens so little that it is hard to detect that an eclipse is in progress. If the Moon grazes the dark central portion of Earth's shadow, the umbra, we see a partial lunar eclipse—the Full Moon looks like it has a dark bite taken from its bright disk.

About every 17 months (on average) the Full Moon passes completely into Earth's shadow. The only sunlight that can reach the Moon during totality comes from red light filtering through Earth's atmosphere, and for up to two hours, the eclipsed Full Moon turns deep red. A total lunar eclipse can be seen from any location on the side of Earth facing toward the Moon.

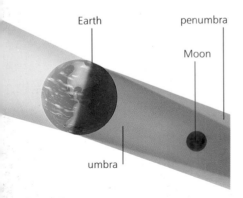

Earth

penumbra

Moon

umbra

LUNAR ECLIPSE When Earth lies between the Moon and the Sun, a lunar eclipse can be seen by everyone on the nighttime side of Earth. In a total eclipse, the entire Moon is in Earth's shadow.

OBSERVING TIPS

The Moon's craters; a dust storm on Mars; Jupiter's Great Red Spot; a galaxy's spiral arms—the night sky offers all this and more to those who know how and where to look.

SOME GENERAL TIPS

Whether you use binoculars or a telescope, make sure the optics are mounted steadily. Telescopes that shimmy and shake cannot show anything clearly.
Seeing The quality of atmospheric seeing determines how much you can see on a given night, so do not observe over rooftops or parking lots, where rising warm air disturbs the view.
Dark-adapted eyes At the beginning of an observing session, take 15 minutes or so to adapt your eyes to the dark. The pupils of your eyes will gradually open to allow in more light from the stars. Use red-filtered flashlights to keep the eyes dark-adapted.

OBSERVING THE MOON

The best time to look for details on the Moon is when sunlight strikes its surface at a shallow

FIRST QUARTER Dark "seas" occupy much of the First Quarter moon. Craters (see close-up at right) are clustered in the south.

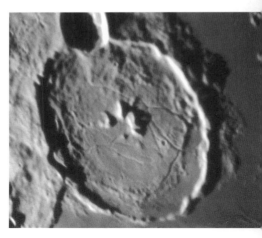

ngle, and the line between lunar night and day throws craters, domes, and hills into starkest relief. This happens twice a month for about a week at a time, centered on the dates of the First and Last Quarter phases. The Full Moon does have its attractions, however. Its flat lighting is best

FULL MOON At Full Moon, light from the Sun flattens all perspective, leaving only variations of light and dark caused by differences in age and composition.

for tracing the bright rays and for examining varied shades that mark different lava flows in the dark "seas," or maria.

Moon-roving Start with the lowest magnification eyepiece your telescope has. Then slowly increase the power as one feature or another catches your eye. If views become unsteady, reduce the magnification until details sharpen again. (Viewing the entire Full Moon is almost painful, so bright is its disk, and some telescopic observers use gray filters to reduce the glare.)

If you have a Moon map, you can start to work your way systematically over the lunar surface. Identify any large maria you can see, and use them as a guide to take you farther.

OBSERVING PLANETS

Planets move slowly among the stars, following patterns that repeat over months or years.

Two inner planets (Mercury and Venus) orbit closer to the Sun than Earth, while six outer planets (Mars through Pluto) orbit farther out than Earth. This difference determines how each planet moves during its apparition, or period of visibility. It also determines how and where you need to look for a planet.

Inner planets Mercury (or Venus) first appears low in the western sky after sunset. It climbs higher day by day, becoming most visible at greatest eastern elongation. It then passes between Earth and the Sun (called inferior conjunction) and moves into the morning sky for best visibility at greatest western elongation. It then moves onward to round the far side of the Sun. This is superior conjunction, the start of a new cycle.

IN THE EYEPIECE
What can you expect to see of the planets? An ice cap on Mars (left) and Jupiter's clouds and Galilean moons (far left) are just some of the features visible in a telescope's eyepiece.

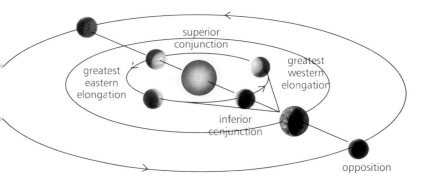

superior
conjunction

greatest
eastern
elongation

greatest
western
elongation

inferior
conjunction

opposition

Outer planets An outer planet's
apparition begins as it becomes
visible low in the east just before
sunrise. Each day the planet rises
earlier. Eventually it rises at
sunset, a point called opposition
which marks best visibility. The
planet then starts setting sooner
and sooner after the Sun, and
finally disappears into the solar
glare. The process restarts when
the planet emerges at dawn again.

Planet-hunting To find a planet,
consult astronomy magazines,
almanacs, or computer software.
Telescopes for planet-watching
need sturdy mountings and high-
quality optics. Most observers use
powers of 200x or less, due to
unsettled seeing. Also useful are
colored filters, which accentuate
specific features on the planets.

INTERIOR–EXTERIOR Interior planets,
such as Venus, reach best visibility at
the points of greatest elongation, while
exterior planets, such as Mars, are best
seen around the time of opposition.

Keep in mind that looking
through a telescope is an acquired
skill. It may take a few months of
viewing before your eyes become
trained to see the finest details on
the small disks of planets.

OBSERVING DEEP-SKY OBJECTS
When we look past the brightest
stars, some of the specks of light
we see are star clusters, nebulas,
and galaxies. Many of these can
be seen using binoculars and
small to medium telescopes. But
no matter what instrument you
use, dark, clear skies are essential

for deep-sky astronomy. Some galaxies are only slightly brighter than the background skyglow, so you need all the contrast you can muster from your instrument and your eyes. Here are a few tools and techniques that can improve your view of these distant objects.

Filters If you live in and around the city, you can effectively boost the contrast by using light pollution reduction (LPR) filters on your telescope. They work by blocking the wavelengths from artificial lighting, and are especially useful for nebulas, many of which are normally invisible from cities.

Averted vision To see faint nebulas and galaxies, try looking off to one side to place the image on the more light-sensitive outer portion of your eye's retina. This technique, called averted vision, will often reveal otherwise invisible deep-sky denizens.

For especially low-contrast objects, use averted vision and the "jiggle" method. Simply give the telescope tube a light tap when you think a nebula is in the field. The eye can detect contrast differences more easily if the image moves slightly.

Magnification What magnification you should use depends on the brightness and size of the object. Objects that have large diameters and faint magnitudes—as many galaxies do—require low magnification and an alertness to low-contrast details, while objects with small diameters and bright magnitudes—such as planetary nebulas—demand greater magnification and accurate positioning of the telescope's field of view.

Filters
LPR filters provide clearer images of some deep-sky objects.

GALAXY CHALLENGE Although bright at magnitude 5.7, the Pinwheel Galaxy is difficult to see because it is spread out over a large patch of sky.

FILTERED VIEW A light-polluted view of the Veil Nebula (above) shows a dramatic improvement when a light pollution filter (LPR) is used (left).

STAR CHARTS

*Road maps to guide you through
the night sky, season by season,
constellation by constellation.*

Seasonal
Star Charts

THESE SEASON-BY-SEASON CHARTS of the night sky
are the essential outdoor guides for identifying the
bright stars and the constellations.

USING THE CHARTS

These maps chart the entire night sky at seasonal intervals. They come in two sets: one for the Northern Hemisphere and one for the Southern.

NORTHERN AND SOUTHERN

The Northern Hemisphere charts (pp. 192–9) are for the latitudes of the United States, Canada, Europe, and Japan. While they show the sky from a latitude of 40°N, the charts are usable for latitudes 10 to 15 degrees north or south of this.

The Southern Hemisphere charts (pp. 200–7) depict the sky from a latitude of 35°S. These are for use in the South Pacific, Australia, New Zealand, South America, and southern Africa.

NORTH AND SOUTH

Each season's sky is divided into two halves—showing the sky as you would see it facing north and facing south. Pick one chart and read the description, then identify one or two bright constellations. After locating one star pattern, use it to jump off to others.

Magnitude
The dimmest stars are magnitude 5.

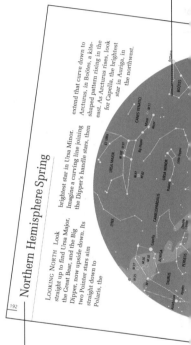

Northern Hemisphere Spring

LOOKING NORTH Look straight up to find Ursa Major, the Great Bear, and the Big Dipper, now upside down. Its two Pointer stars aim straight down to Polaris, the brightest star in Ursa Minor. Imagine a curving line joining the Dipper's handle stars, then extend that curve down to Arcturus, in Boötes, a kite-shaped pattern rising in the east. As Arcturus rises, look for Capella, the brightest star in Auriga, in the northwest.

192

Hemisphere
Select the hemisphere you live in, then the current season. Each pair of maps for that season depicts the night sky you will see as you face north or south.

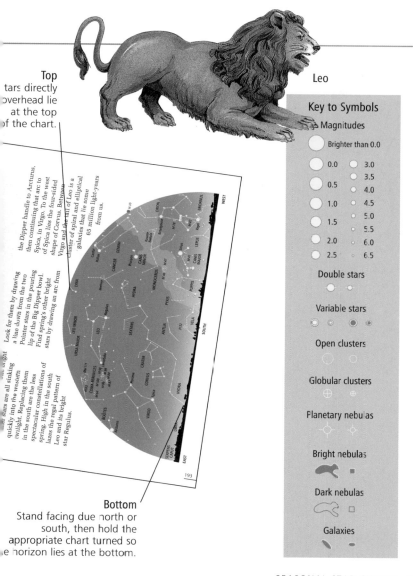

Top
Stars directly overhead lie at the top of the chart.

Leo

Look for them by drawing a line down from the two Pointer stars in the pouring lip of the Big Dipper bowl. Find spring's other bright stars by drawing an arc from the Dipper handle to Arcturus, then continuing that arc to Spica, in Virgo. To the west of Spica lies the four-sided shape of Corvus. Between Virgo and the tail of Leo is a cluster of spiral and elliptical galaxies that lie some 65 million light-years from us.

...stars are all sinking quickly into the western twilight. Replacing them in the south are the less spectacular constellations of spring. High in the south blazes the regal pattern of Leo and its bright star Regulus.

Bottom
Stand facing due north or south, then hold the appropriate chart turned so the horizon lies at the bottom.

Key to Symbols

Magnitudes

Brighter than 0.0

0.0	3.0
0.5	3.5
1.0	4.0
1.5	4.5
2.0	5.0
2.5	5.5
	6.0
	6.5

Double stars

Variable stars

Open clusters

Globular clusters

Planetary nebulas

Bright nebulas

Dark nebulas

Galaxies

Northern Hemisphere Spring

LOOKING NORTH Look straight up to find Ursa Major, the Great Bear, and the Big Dipper, now upside down. Its two Pointer stars aim straight down to Polaris, the brightest star in Ursa Minor. Imagine a curving line joining the Dipper's handle stars, then extend that curve down to Arcturus, in Boötes, a kite-shaped pattern rising in the east. As Arcturus rises, look for Capella, the brightest star in Auriga, in the northwest.

LOOKING SOUTH The bright winter stars are all sinking quickly into the western twilight. Replacing them in the south are the less spectacular constellations of spring. High in the south lazes the regal pattern of Leo and its bright star Regulus.

Look for them by drawing a line down from the two Pointer stars in the pouring lip of the Big Dipper bowl. Find spring's other bright stars by drawing an arc from the Dipper handle to Arcturus, then continuing that arc to Spica, in Virgo. To the west of Spica lies the four-sided shape of Corvus. Between Virgo and the tail of Leo is a cluster of spiral and elliptical galaxies that lie some 65 million light-years from us.

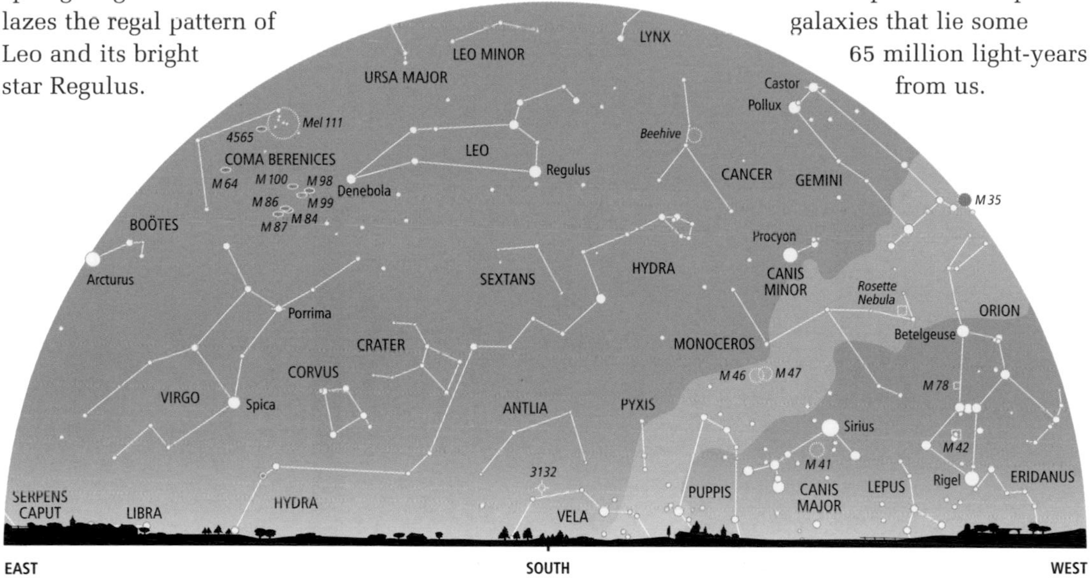

EAST

SOUTH

WEST

193

Northern Hemisphere Summer

LOOKING NORTH During summer, the turning of the sky around Polaris sinks the Big Dipper into the northwest, swings the head of Draco overhead, and raises Cassiopeia into the northeast. In the east, Vega, in Lyra, and Deneb, in Cygnus, form a large figure in the sky with Altair, in Aquila, which is found on the southern map (see facing page). These three stars, the brightest of summer, make up the Summer Triangle.

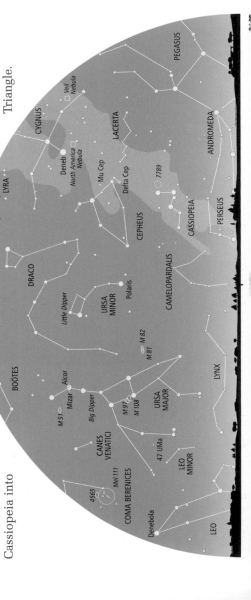

LOOKING SOUTH Straight overhead, look for the crooked "H" pattern of Hercules. Arcturus and the sparse stars of spring still dominate the western sky. Due south, look for Antares amid the hook-shaped pattern of Scorpius. To the left of Scorpius,

a teapot made of stars marks Sagittarius. In this direction you are looking toward the center of our Galaxy, which is hidden behind 25,000 light-years of dusty gas. If the sky

above your observing site is very dark (with no Moon) and haze-free, you will see what looks like steam rising from the teapot and wafting up the sky past the bright stars Altair and Vega. That "steam" is the Milky Way, made of millions of distant stars.

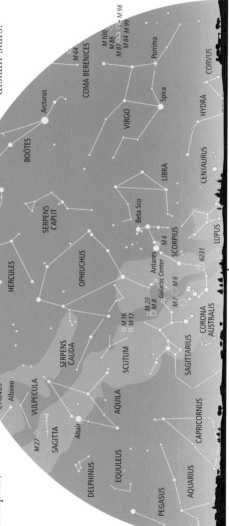

Northern Hemisphere Autumn

LOOKING NORTH The Big Dipper skims the horizon, while to the right of Polaris, Cassiopeia climbs into the northeast. To the right of Cassiopeia shines Andromeda, her daughter in mythology, and Perseus, the hero who rescued Andromeda from Cetus the Sea Monster, itself visible low in the south. Of the bright stars, Deneb shines down from directly overhead, Arcturus is setting in the northwest, and Capella is rising in the northeast.

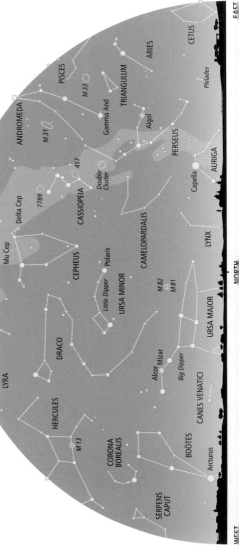

LOOKING SOUTH Sagittarius and the center of the Galaxy sink into the southwest. Due south, look for a chevron of faint stars that marks Capricornus. Below shines a lone bright star – Fomalhaut, in Piscis Austrinus. The Summer Triangle stars of Altair, Deneb, and Vega (the latter two on the northern chart, see facing page) now stand high in the west. They will continue to dominate the western sky through the season. However, the true autumn constellations gradually take the scene, led by Pegasus (marked by a square of four stars) rising into the east. The Flying Horse is upside down, so the line of stars to the lower right of the square delineates his neck.

EAST

SOUTH

WEST

HERCULES

SERPENS CAPUT

OPHIUCHUS

M16
M17

SERPENS CAUDA

M20
M8

Galactic Center

SCUTUM

LYRA

M57

VULPECULA

Albireo

M27

SAGITTA

SAGITTARIUS

Altair

AQUILA

CORONA AUSTRALIS

DELPHINUS

CAPRICORNUS

CYGNUS

Veil Nebula

EQUULEUS

LACERTA

AQUARIUS

MICROSCOPIUM

INDUS

PEGASUS

PISCIS AUSTRINUS

Fomalhaut

GRUS

SCULPTOR

PISCES

CETUS

Mira

Northern Hemisphere Winter

LOOKING NORTH The Big Dipper, within Ursa Major, stands on its handle low on the northeast horizon. Meanwhile, the Little Dipper hangs by its handle from Polaris, the North Star. As you face north, Cassiopeia's "W" shape of five stars now looks more like a celestial letter M. The twin stars of Castor and Pollux in Gemini are rising in the eastern sky while Capella, in Auriga, shines from high in the northeast.

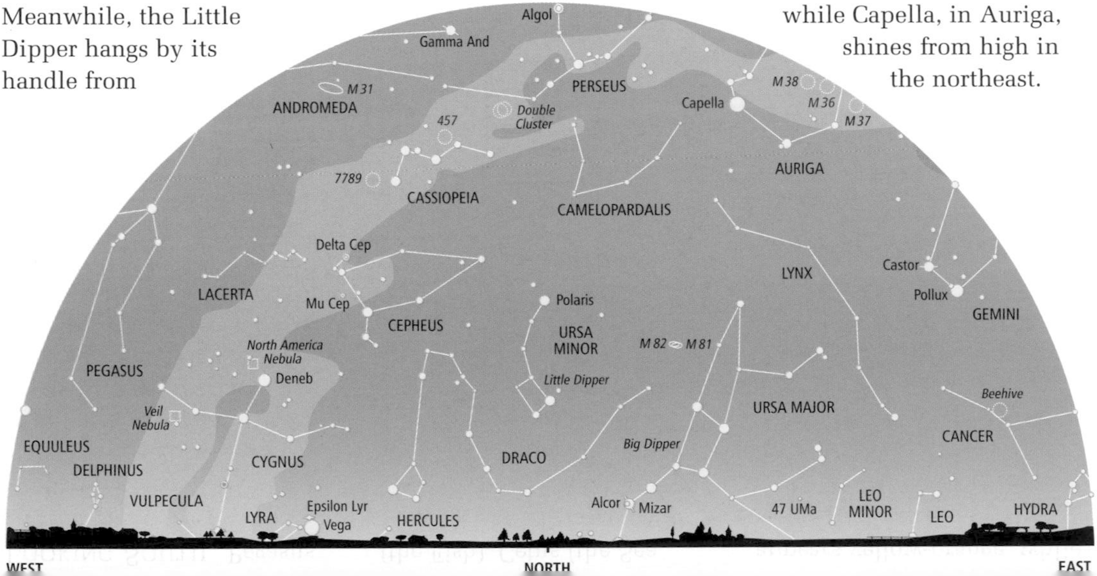

EAST

NORTH

WEST

HYDRA
GEMINI
Pollux
Castor
CANCER
Beehive
LEO
LEO MINOR
AURIGA
M37
M36
M38
Capella
LYNX
URSA MAJOR
47 UMa
M82 M81
Big Dipper
Mizar
Alcor
CAMELOPARDALIS
PERSEUS
Algol
Double Cluster
Gamma And
457
CASSIOPEIA
Polaris
URSA MINOR
Little Dipper
DRACO
CEPHEUS
HERCULES
M31
ANDROMEDA
7789
Delta Cep
Mu Cep
North America Nebula
Deneb
Epsilon Lyr
Vega
LYRA
CYGNUS
LACERTA
Veil Nebula
VULPECULA
PEGASUS
DELPHINUS
EQUULEUS

LOOKING SOUTH Pegasus sinks into the west, followed by an arc of stars that marks Andromeda, Cassiopeia's mythological daughter. Her hero, Perseus, swings high overhead. A chain of dim, "watery" constellations—Aquarius (the Water Carrier), Pisces (the Fish), Cetus (the Sea Monster), and Eridanus (the River)—sprawl in the south-west. By contrast, bright stars surround Orion, climbing into the east. Aldebaran in Taurus appears yellow-orange, while above sparkles the lovely Pleiades star cluster. (Taurus's face is formed from a looser star cluster, the Hyades.) The sky's brightest star, Sirius, in Canis Major, sparkles in the southeast; blue-white Procyon shines above it.

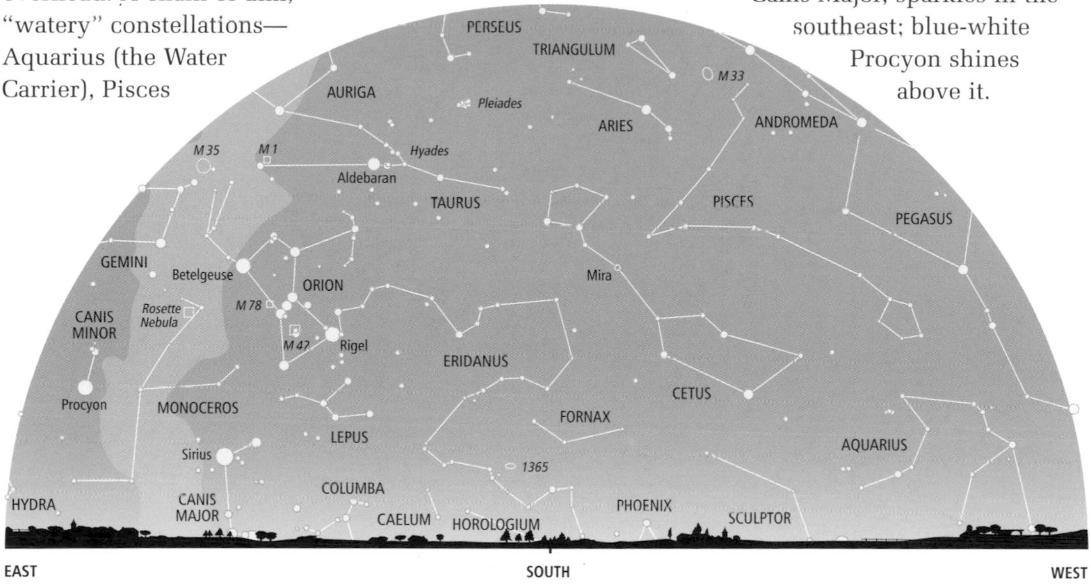

Southern Hemisphere Spring

LOOKING NORTH   The Milky Way edges westward but is still prominent as it cascades from the star clouds of Sagittarius to flow between the triangle formed by the stars Altair, Vega, and Deneb, just skimming the northern horizon. As the Milky Way retreats to the west, new spring constellations appear in the east. The most notable is the square of four stars that marks Pegasus. High in the north-east is Fomalhaut, in Piscis Austrinus.

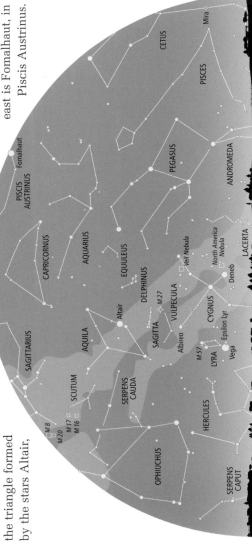

LOOKING SOUTH As spring takes hold, Crux, and Alpha and Beta Centauri, drop low into the southwest, with Scorpius and its rich star clouds following not far behind. The departure of the winter Milky Way leaves a spring sky populated with faint constellations and few bright stars. High overhead, look for Grus, one of the few constellations in this sky with an easy-to-identify pattern. Below Grus shines bright Achernar in Eridanus. Just to the right, or west, is the Small Magellanic Cloud (SMC), now climbing into prominence. Directly below, about halfway between the SMC and the horizon, lies another satellite galaxy to our own: the Large Magellanic Cloud (LMC).

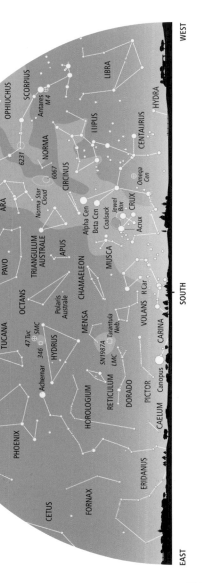

Southern Hemisphere Summer

LOOKING NORTH Orion stands on his head due north. Look for Y-shaped Lepus at his feet. A line drawn through the stars of Orion's Belt and extended up to the right points to

Sirius, the "dog star." The same Belt line extended down points to ruddy Aldebaran, the

eye of Taurus. Below it is the Pleiades star cluster. Capella twinkles just above the northern horizon, with the twin stars of Castor and Pollux rising to the right, or east.

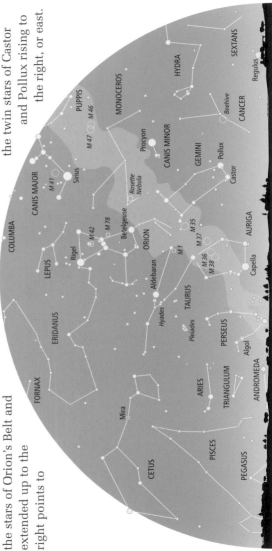

LOOKING SOUTH The galaxies known as the Large and Small Magellanic Clouds (LMC and SMC) appear high in the south, as obvious hazy patches amid the faint starfields of Tucana, Hydrus, and Dorado. The brightest star in the southwest is Achernar, at one end of Eridanus. Canopus, the night sky's second brightest star, blazes above the LMC. To the left, or east, of the clouds, the Milky Way tumbles out of the northern sky's Puppis, down through Vela and Carina, then on past Crux. Puppis (the Stern), Vela (the Sails), and Carina (the Keel), along with Pyxis (the Compass), once formed a constellation called Argo Navis—the Ship Argo—the vessel of Jason and the Argonauts.

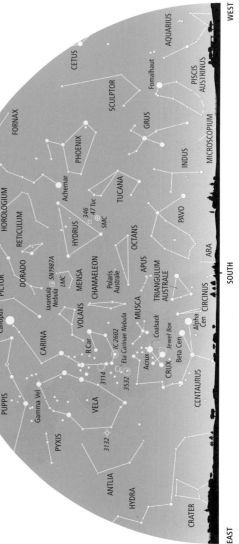

EAST

SOUTH

WEST

203

Southern Hemisphere Autumn

LOOKING NORTH Orion and the bright stars of the northern summer sky are sinking into the west. Taking their place is a fairly sparse area of sky. Leo, the most distinctive constellation,

is distinguished by Regulus, a beacon in the northeast. A hook of stars hanging from

Regulus marks Leo's head and mane (like many older constellations invented in the Northern Hemisphere, Leo appears upside down when viewed from southern latitudes).

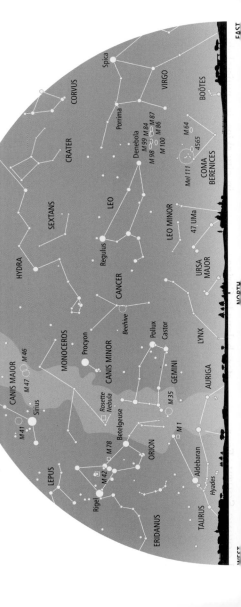

LOOKING SOUTH Four bright stars form the pattern known as the False Cross, which stands due south at this time of year. Though distinctive, it is not a proper constellation. The False Cross's two top stars belong to Vela, while its bottom two stars belong to Carina. Canopus, the brilliant star far to the right, or west, also belongs to Carina. Crux, the real Southern Cross, its stars bright enough to be seen even in city skies, is unmistakable, high in the southeast, as are nearby Beta and Alpha Centauri, which delineate the forefeet of Centaurus, the centaur of Greek mythology. These two stars are also known as the Pointers, because they point directly to Crux.

EAST

SOUTH

WEST

VIRGO

LIBRA

SCORPIUS

LUPUS

HYDRA

CENTAURUS

Centaurus A
Omega Cen

CRUX

Acrux
Jewel Box
Coalsack

Eta Carinae Nebula

3532

3132

ANTLIA

VELA

Gamma Vel

False Cross

3114

IC 2602

R Car

CARINA

PYXIS

PUPPIS

Canopus

NORMA

6067
Norma
Star Cloud

CIRCINUS

Alpha Cen

Beta Cen

MUSCA

APUS

CHAMAELEON

VOLANS

Tarantula
Nebula

SN1987A
LMC

MENSA

Polaris
Australe

OCTANS

PICTOR

COLUMBA

DORADO

CAELUM

RETICULUM

HOROLOGIUM

SMC
47 Tuc
316
HYDRUS

IULANA

INDUS

PAVO

ARA

TRIANGULUM
AUSTRALE

Achernar

PHOENIX

FORNAX

ERIDANUS

1365

LEPUS

205

Southern Hemisphere Winter

LOOKING NORTH The brightest star in the north is Arcturus, in Boötes, the fourth brightest in the sky. To the right, or east, of Arcturus a neat semi-circle of stars forms Corona Borealis. Above Arcturus shines fainter Spica, in Virgo, while high overhead ruddy Antares gleams in Scorpius. From Scorpius the Milky Way flows down into the northeastern sky through the constellations of Sagittarius, Scutum, and Aquila.

LOOKING SOUTH The Milky Way and the easy-to-identify constellations of Crux and Centaurus begin their descent into the west. Rising high into the east is the "fish-hook" of stars that forms Scorpius, followed by a "teapot" of stars that marks

Sagittarius. If possible, on the next moonless night, get away from city lights and explore the Milky Way with a pair of binoculars. Among other things, look for the shape of

a giant "emu" formed by dark lanes in the Milky Way. The Coalsack, a dark nebula near Crux, forms his head; a dark band that stretches from Alpha Centauri to Scorpius forms his neck; and his legs are found in dark lanes streaming down from Scorpius.

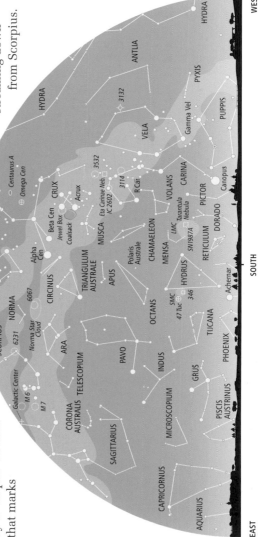

EAST

SOUTH

WEST

Constellation Charts

YOU ARE ABOUT TO EMBARK on a series of guided tours
of the cosmos, constellation by constellation. The
clear directions and easy-to-use star charts will show
you a treasure trove of celestial objects.

USING THE CHARTS

After some practice with the seasonal charts you will probably want to zero in on particular constellations.

THE A–Z OF CONSTELLATIONS

For ease of use, the charts (pp. 212–307) are presented in alphabetical order. North is at the top, and east is at the *left*—different from Earth maps, but necessary to match our view of the sky. All stars marked are visible to the naked eye, but near cities, binoculars may be needed to pick out the fainter ones.

Instruments Many of the deep-sky objects noted are within the range of binoculars or telescopes with apertures of 2.4 inches (60 mm). Of course most objects will reveal more detail when viewed with a larger telescope.

Magnitude
The dimmest stars are magnitude 6.5.

Auriga

oh-RYE-gah

The Charioteer

This lovely multi-sided figure is easy to find in the sky, largely because of bright Capella. Ancient legends portray Auriga as a chario-teer carrying a goat on his shoulder.

◈ EPSILON (ε) AURIGAE An extraordinary variable system, this supergiant star fades when its com-panion passes in front of it once every 27 years. During an eclipse, its brightness drops by two-thirds of a magnitude.

220

M 37 This is an

M 36 This has

Deep-sky objects
Many deep-sky objec visible in a small telescope are plotted

CHART SYMBOLS

Magnitudes ⬤ -1 ⬤ 0 ⬤ 1 ● 2 ● 3 • 4 · 5 · 6 and under

Double stars •– Variable stars ⊙⊙ Open clusters ◌◌

Globular clusters ⊕⊖ Planetary nebulas ◇◇

Diffuse nebulas ◁□▫ Galaxies ◯◦ Quasar ⊚

Information panel
The panel shows the constellation's genitive name, abbreviation, and best viewing time.

BOÖTES (BOO)
On meridian
10 p.m. June 1

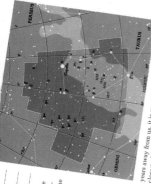

BOÖTES (BOO)
On meridian
10 p.m. June 1

Boötes

boh-OH-teez

The Herdsman

In one Greek myth, Boötes was the son of Demeter, goddess of agriculture. He is said to have invented the plow, and as a reward was placed in the sky by the gods. In another legend, Boötes was the son of Zeus and his mistress, Callisto. Callisto, jealous wife of Zeus, was almost killed by her son Boötes when he was out hunting. Zeus rescued her, became the constellation Ursa Major, the Great Bear.

years away from us, it is one of the closest of the bright stars. Arcturus's actual position in the sky has changed by more than twice the Moon's apparent diameter in the last 2,000 years. Astronomers say that Arcturus has a large proper motion.

ARCTURUS (ALPHA [α] BOÖTIS). This yellow-orange star is the fourth brightest star in the sky, and is easily identified. At 37 light-

221

Description
Refer to the description for highlights of what to look for in a particular constellation.

Icons
These show the "tool" needed to see the object. See key below.

 Naked eye symbol

Binoculars symbol

Telescope symbol

Andromeda

an-DROH-me-duh

The Chained Princess

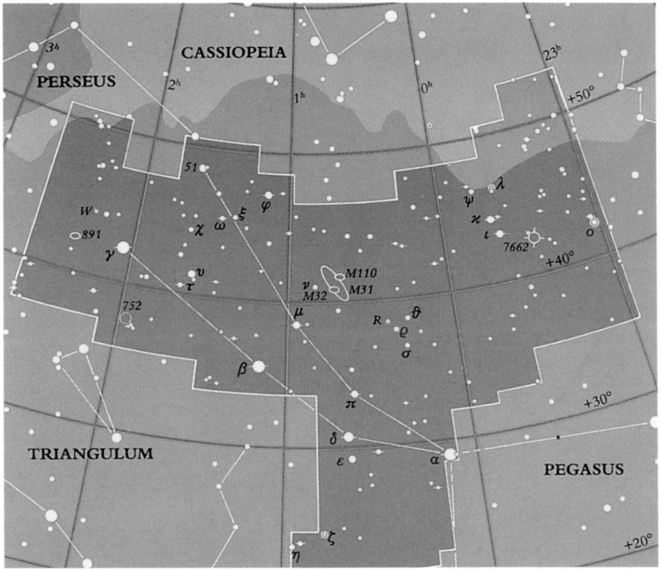

Although Andromeda is renowned for the great galaxy that resides within the constellation, its stars are not very bright. It is easy to find, however, located south of Cassiopeia's W, and just off one corner of the Great Square of Pegasus.

THE ANDROMEDA GALAXY (M 31) The closest major galaxy to us, the Andromeda Galaxy was first thought to be a nebula, and was listed in comet hunter Charles Messier's 18th-century catalog of nebulas. A spiral galaxy much like our own Milky Way, it is a maelstrom comprising 200 billion stars and clouds of dust and gas. It is bright enough to be seen with binoculars from city sites and with the naked eye beneath a dark sky, being one of the most distant objects visible to the unaided eye. In the field of larger binoculars, or using a small telescope, you can see its two neighboring elliptical galaxies,

ANDROMEDAE (AND)
On meridian
10 p.m. Nov 1

M 32 and M 110. M 32 is small and compact; M 110 is larger and more diffuse, and is thus harder to see.

 GAMMA (γ) ANDROMEDAE The brighter member of this double is a golden yellow, and its companion is greenish blue.

 R ANDROMEDAE This long-period Mira variable has a range of nine magnitudes.

 NGC 752 This open cluster lies about 5 degrees south of Gamma (γ) Andromedae and is easy to find because of its relatively bright stars. Because it is spread out over such a large area, it is actually easier to see through binoculars than through a telescope. If using a telescope, use it at its lowest power.

 NGC 7662 This fairly bright planetary nebula looks starlike through the smallest telescopes. However, through a 6 inch (150 mm) telescope at moderate power, it becomes a graceful, glowing spot of gas about 30 arcseconds across.

NGC 891 This is a challenge even for 6 inch (150 mm) telescopes, but with a dark sky, you will see one of the best examples of a spiral galaxy, viewed edge-on.

ANDROMEDA GALAXY

With a telescope, use the lowest power you have to fit as much of Andromeda in the field of view as possible. Two companion galaxies are visible in this photo: M 110 (lower right) and M 32, the starlike blob left of Andromeda's center.

Antlia

ANT-lee-uh

The Air Pump

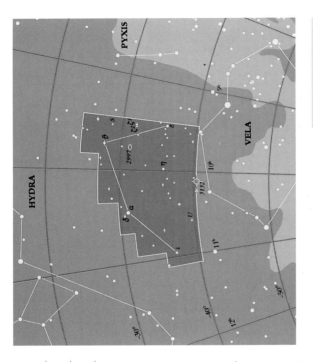

The Air Pump was named after the 17th-century physicist Robert Boyle's invention by Nicolas-Louis de Lacaille during the time he spent working at an observatory at the Cape of Good Hope, from 1750 to 1754. Antlia is a small, faint constellation just off the bright southern Milky Way, not far from Vela and Puppis. Its alpha (α) star is barely the constellation's brightest star and has been given no proper name. It is red in color and possibly varies slightly in magnitude.

NGC 2997 This is a large, faint spiral galaxy, with a star-like nucleus. It is quite difficult to observe with a small telescope.

Apus

ay-pus

The Bird of Paradise

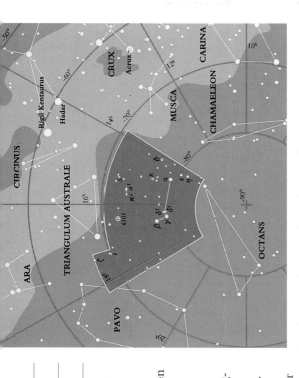

APODIS (APS)
On meridian
10 p.m. June 20

This faint constellation is directly below Triangulum Australe, the Southern Triangle. Being close to the southern pole, it cannot be seen from most northern latitudes.

THETA (θ) APODIS This variable star ranges from magnitude 6.4 to below 8 in a semi-regular cycle over a hundred days.

NGC 6101 A faint globular cluster, NGC 6101 can be seen as a small, misty spot through a small telescope.

S APODIS This is a "backward" nova. Usually the star shines at about magnitude 10—

bright enough to see through a small telescope—but at irregular intervals it erupts dark material. It then fades by about a hundred times, to around magnitude 15. After staying faint for several weeks, it slowly returns to its original brightness.

215

Aquarius

ah-KWAIR-ee-us

The Water Bearer

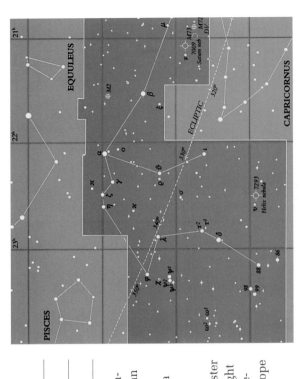

This is one of the sky's oldest constellations, dating from Babylonian times. Appropriately, the Water Bearer is not far from a dolphin, a river, a sea serpent, and a fish.

M 2 This fine globular cluster appears as a fuzzy spot of light through binoculars and small telescopes. A 6 inch (150 mm) telescope resolves it into stars.

THE SATURN NEBULA (NGC 7009) This planetary nebula was so named because its protruding rays made it look like a dim version of Saturn with its rings. It is visible through a telescope as a greenish point of light.

THE HELIX NEBULA (NGC 7293) The Helix is a large planetary nebula. Because its brightness is spread over a large area, it appears best with a low-power, wide-field telescope or binoculars under a dark sky.

Aquila

uh-KWI-luh

The Eagle

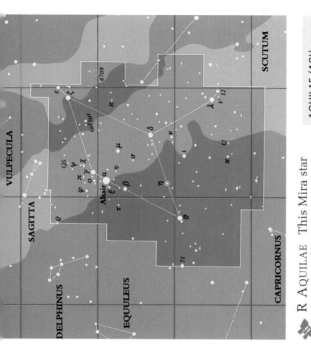

Likened to an eagle by the sky-watchers of the Euphrates Basin, the constellation of Aquila takes its name from the bird that belonged to the Greek god Zeus. Aquila's main accomplishment was to bring the handsome mortal youth Ganymede to the sky to serve as his master's cup bearer.

👁 ETA (η) AQUILAE This supergiant star is a bright Cepheid variable that changes by about a magnitude in brightness (3.5 to 4.4) in a period of little more than a week. At its brightest, it rivals the nearby Delta (δ) Aquilae, and it fades to about the magnitude of Iota (ι) Aquilae.

AQUILAE (AQL)
On meridian
10 p.m. Aug 10

🔭 R AQUILAE This Mira star varies in magnitude from 6 to 11.5 over a period of 284 days.

🔭 NGC 6709 This open cluster is a closely knit group against a rich background of stars.

217

Ara

AR-ah

The Altar

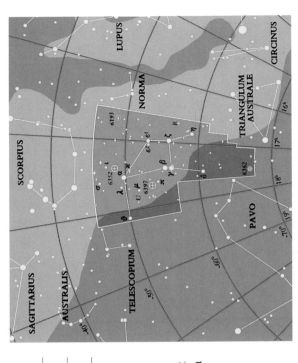

ARAE (ARA)
On meridian
10 p.m. July 10

Located south of Scorpius, Ara's original Latin name was Ara Centauri—the altar of the centaur Chiron. Half man and half horse, Chiron was thought to be the wisest creature on Earth. Ara has also been referred to variously as the altar of Dionysus; the altar built by Noah after the flood; the altar built by Moses; and even the altar from Solomon's Temple.

NGC 6397 This globular cluster is placed between Beta (β) Arae and Theta (ϑ) Arae. It is relatively loose, so an observer with powerful binoculars should be able to detect it without difficulty and perhaps even resolve its faint stars.

U ARAE This Mira-type variable is bright enough to be seen through a small telescope when it is at its maximum of magnitude 8. However, it then drops a full five magnitudes before rising again over a period of more than seven months.

Aries

AIR-eez

The Ram

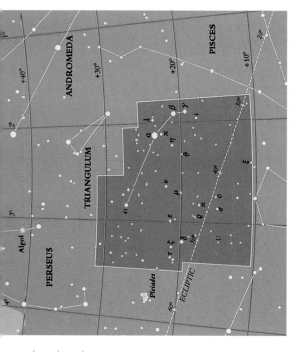

Algol

PERSEUS

ANDROMEDA

TRIANGULUM

PISCES

Pleiades

ECLIPTIC

ARIETIS (ARI)
On meridian
10 p.m. Nov 20

Aries is well known and is not hard to find, but it is small and contains few objects of interest.

The constellation has long been seen as a ram. For the Greeks, it represented the ram from which the golden fleece was taken. In one version of the legend, the god Hermes sent a ram with a golden fleece to carry Phrixus and Helle, the two children of the king of Thessaly, to safety from their cruel stepmother. Helle fell from the ram's back, but Phrixus was carried to safety on the shores of the Black Sea. Here, he sacrificed the ram, and its fleece was placed in the care of a sleepless dragon. It was from here that Jason and the Argonauts stole it.

GAMMA (γ) ARIETIS In 1664, the English scientist Robert Hooke was following the motion of a comet when he chanced upon this beautiful double star. It has a separation of 8 arcseconds, and is easy to find and observe.

Auriga

oh-RYE-gah

The Charioteer

This lovely multi-sided figure is easy to find in the sky, largely because of bright Capella. Ancient legends portray Auriga as a charioteer carrying a goat on his shoulder.

◈ EPSILON (ε) AURIGAE An extraordinary variable system, this supergiant star fades when its companion passes in front of it once every 27 years. During an eclipse, its brightness drops by two-thirds of a magnitude.

🔭 M 36 This bright open star cluster, some 5 degrees south-west of Theta (ϑ) Aurigae, contains about 60 stars of 8th magnitude and fainter.

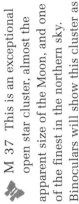

🔭 M 37 This is an exceptional open star cluster, almost the apparent size of the Moon, and one of the finest in the northern sky. Binoculars will show this cluster as a misty spot. A small telescope will reveal its large number of stars.

220

Boötes

boh-OH-teez

The Herdsman

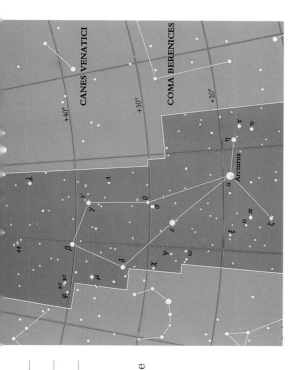

BOÖTIS (BOO)
On meridian
10 p.m. June 1

In one Greek myth, Boötes was the son of Demeter, goddess of agriculture. He is said to have invented the plow, and as a reward was placed in the sky by the gods. In another legend, Boötes was the son of Zeus and his mistress, Callisto. Callisto, changed into a bear by Hera, the jealous wife of Zeus, was almost killed by her son Boötes when he was out hunting. Zeus rescued her, taking her into the sky where she became the constellation Ursa Major, the Great Bear.

🔍 ARCTURUS (ALPHA [α] BOÖTIS) This yellow-orange star is the fourth brightest star in the sky, and is easily identified. At 37 light-years away from us, it is one of the closest of the bright stars. Arcturus's actual position in the sky has changed by more than twice the Moon's apparent diameter in the last 2,000 years. Astronomers say that Arcturus has a large proper motion.

221

Caelum

SEE-lum

The Chisel

Caelum is one of the least conspicuous of all the constellations, containing a few faint stars with a magnitude of 5 at best. Originally named Caela Sculptoris, the Sculptor's tool, it is one of the many regions in the Southern Hemisphere skies that was named by the 18th-century astronomer Nicolas-Louis de Lacaille. It comprises a largely empty region of the heavens between the constellations of Columba, the Dove, and Eridanus, the River.

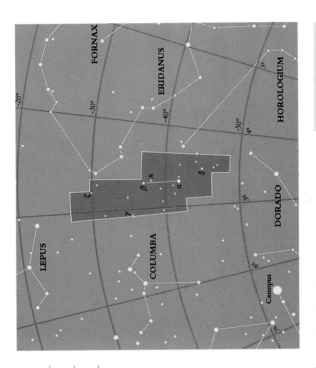

CAELI (CAEI)
On meridian
10 p.m. Jan 1

R CAELI A bright Mira-type variable, this star changes from magnitude 6.7 to 13.7 over a period of about 13 months.

Camelopardalis

ka-mel-o-PAR-da-lis

The Giraffe

Camelopardalis was dreamed up by the German astronomer Jakob Bartsch in 1624, who claimed that it represented the camel that brought Rebecca to Isaac. The constellation lies between Auriga and the two bears.

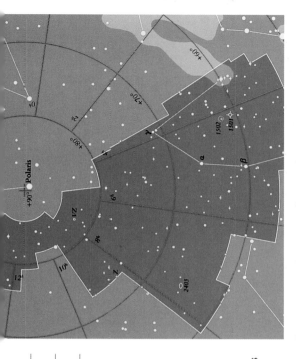

CAMELOPARDALIS (CAM)
On meridian
10 p.m. Jan 10

Z CAMELOPARDALIS This cataclysmic variable star erupts every two or three weeks from its minimum of magnitude 13 to a maximum of 9.6, which is still quite faint. Its resemblance to other such variables ceases when, while fading, it stops changing and hovers at an intermediate magnitude. This "standstill" might last for months before the decline resumes. In the late 1970s, Z Cam stayed around magnitude 11.7 for several years.

VZ CAMELOPARDALIS This star varies irregularly over the small range between magnitudes 4.8 and 5.2. It is located close to Polaris.

Cancer

CAN-ser

The Crab

Cancer is the faintest member of the zodiac, and its main claim to fame is the beautiful M 44.

THE PRAESEPE OR BEEHIVE (M 44) One of the sky's finest open clusters, this is easy to see through binoculars from the city and with the naked eye from a dark site. There are over 200 stars in the Praesepe. Spread over 1½ degrees, they are best seen with binoculars.

M 67 This open cluster has 500 faint stars spread over half a degree. Although you can find it with binoculars, your best view will be through a small telescope's low-power eyepiece.

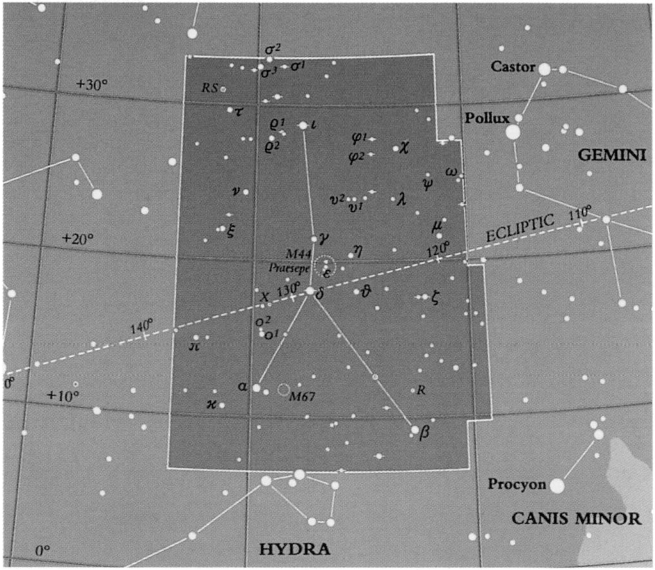

R CANCRI This bright long-period variable is easily visible through a pair of binoculars when near its maximum magnitude of 6.2. R Cancri varies down to a faint 11.2 and back in almost precisely a year.

CANCRI (CNC)
On meridian
10 p.m. March 1

Canes Venatici

KAH-nez ve-NAT-eh-see

The Hunting Dogs

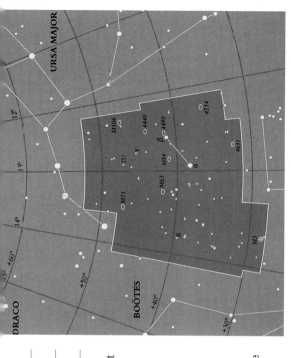

This constellation, tucked away just south of the Big Dipper's handle, contains a wide variety of deep-sky objects.

◆ COR CAROLI The heart of Charles, Alpha (α) Canum Venaticorum is believed to have been named by Edmond Halley after his patron, Charles II. This wide double is easily split.

✹ M 3 A rare gem of the northern sky, this globular cluster is midway between Cor Caroli and Arcturus. Some 35,000 light-years away and 200 light-years across, M 3 begins to resolve into stars through a small telescope.

CANUM
VENATICORUM
(CVN)
On meridian
10 p.m. May 1

✹ THE WHIRLPOOL GALAXY (M 51) In binoculars, this galaxy appears as a round, 8th-magnitude glow with a bright nucleus. A 12 inch (300 mm) telescope will show its spiral structure. M 51 lies 23 million light-years away.

225

Canis Major

KAH-niss MAY-jer

The Great Dog

One of the most striking of all the constellations, the Great Dog is marked by the brilliant star Sirius, commonly known as the Dog Star—the brightest star in the entire sky.

Canis Major and its neighboring constellation, Canis Minor, the Little Dog, appear in a number of myths. One legend has the two dogs sitting patiently under a table at which the Twins are dining. The faint stars that can be seen scattered in the sky between Canis Minor and Gemini are the crumbs the Twins have been feeding to the animals.

According to the ancient Greeks, Canis Major could run incredibly fast. Laelaps, as they called him, is said to have won a race against a fox

that was the fastest creature in the world. Zeus placed the dog in the sky to celebrate the victory. Another myth has the Great Dog and Little Dog assisting Orion while he is hunting. In other versions of the story, Sirius is Orion's hunting dog.

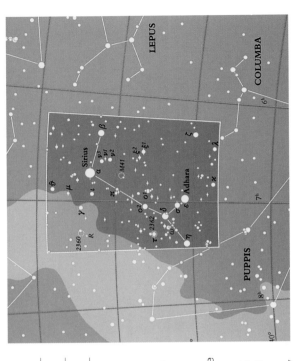

CANIS MAJORIS
(CMA)
On meridian
10 p.m. Feb 1

For the ancient Egyptians, Sirius had special meaning. After being close to the Sun for several months, the star would rise just before dawn in late summer. This event heralded the annual flooding of the Nile Valley, the waters refertilizing the fields with silt.

◈ SIRIUS The sky's brightest star, Sirius is only 8.6 light-years from Earth. Its great brilliance is also due to its being some 40 times more luminous than the Sun. In 1834, Friedrich Bessel noted that Sirius had a strange wobble to its position, indicating an unseen companion. In 1862, the famous telescope maker Alvan Clark, while testing a new 18½ inch (460 mm) refractor on Sirius, discovered the faint star we now know as the Pup. It is a white dwarf star, its density being so great that a piece of it the size of this book might weigh.

around 200 tons (203 tonnes). On its own, the Pup would be a respectable star visible through a telescope at magnitude 8.4, but its closeness to mighty Sirius makes it a very difficult target, requiring a telescope of 10 inch (250 mm) aperture and very steady viewing conditions.

✦ M 41 A beautiful open cluster, M 41 is surrounded by a rich field of background stars. If you look at it through a telescope, you will be able to see a distinctly red star near the cluster's center.

✦ NGC 2362 This cluster of several dozen stars is tightly packed around Tau (τ) Canis Majoris. What is not clear is whether Tau (τ) is actually a member of the cluster or just a chance foreground star.

THE GREAT DOG

Canis Major has been depicted as a dog for thousands of years. The constellation is easily visible from most parts of Earth.

Canis Minor

KAH-niss MY-ner

The Little Dog

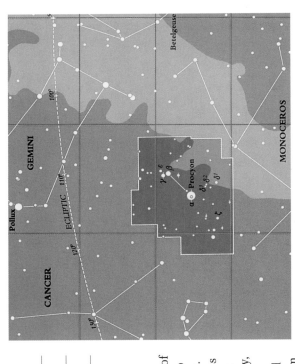

Canis Minor has only two stars brighter than 5th magnitude—Procyon and Gomeisa (Beta [β] Canis Minoris). Besides being one of Orion's dogs, Canis Minor was also said to be one of Actaeon's hounds. Actaeon surprised Artemis, goddess of the hunt, while she was bathing outdoors. Spellbound by her beauty, he paused for a moment and she saw him. Furious that a mortal had seen her naked, Artemis turned him into a stag, set her hounds upon him, and he was devoured.

CANIS MINORIS
(CMI)
On meridian
10 p.m. Feb 10

👁 PROCYON (ALPHA [α] CANIS MINORIS) This beautiful deep yellow star follows Orion across the sky. Only 11.4 light-years away, it is accompanied by a white dwarf that is much fainter than the Pup that accompanies Sirius.

🔭 BETA (β) CANIS MINORIS This is set in a beautiful field that includes one quite red star.

Capricornus

kap-reh-KOR-nuss

The Sea Goat; Capricorn

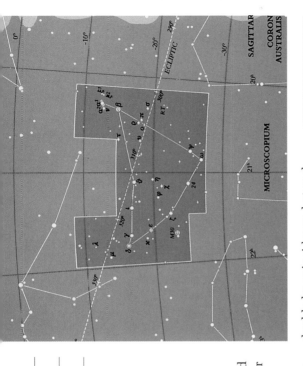

The triangle of stars forming Capricornus is easily recognized, although the stars are no brighter than 3rd magnitude.

The constellation has been named for a goat since Babylonian times, and it is often depicted as a goat with the tail of a fish. This might relate to a story about the god Pan, who, when fleeing the monster Typhon, leaped into the Nile. The part of him that was underwater turned into a fish tail, while his top half remained that of a goat.

ALPHA (α) CAPRICORNI This double has a separation of 6 arcminutes—a naked-eye test for a night's clarity and steadiness. The pair is a double by coincidence, but each star is itself a true binary.

M 30 Perhaps 40,000 light-years away, this globular cluster has a fairly dense center. It is not well resolved in small telescopes.

CAPRICORNI (CAP)
On meridian
10 p.m. Sept 1

Carina

ka-RYE-nah

The Keel

This constellation lies in one of the richest parts of the Milky Way, and under a dark sky it is breathtaking.

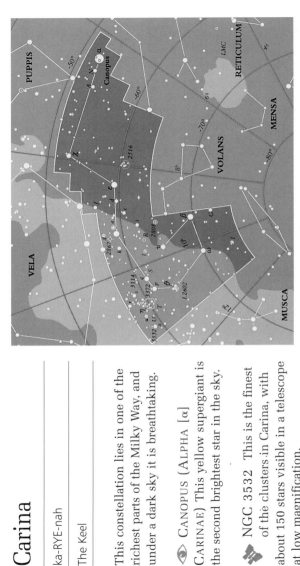

CANOPUS (ALPHA [α] CARINAE) This yellow supergiant is the second brightest star in the sky.

NGC 3532 This is the finest of the clusters in Carina, with about 150 stars visible in a telescope at low magnification.

THE ETA (η) CARINAE NEBULA (NGC 3372) This exquisite nebula will reward any size of binoculars or telescope. It is 2 degrees across, with dark rifts appearing to break it up. The dark

Keyhole Nebula (NGC 3324) is superimposed on its brightest part.

IC 2602 This open cluster around Theta (θ) Carinae is best seen in binoculars or in the eyepiece of a wide-field telescope.

CARINAE (CAR)
On meridian
10 p.m. March 1

Cassiopeia

kass-ee-oh-PEE-uh

The Queen

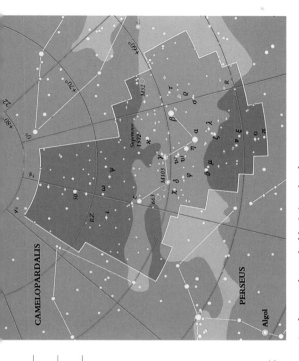

CASSIOPEIAE (CAS)
On meridian
10 p.m. Nov 1

This striking W-shaped figure is on the other side of Polaris from the Big Dipper. Most prominent in the Northern Hemisphere's winter sky, Cassiopeia is visible all year from mid-northern latitudes.

GAMMA (γ) CASSIOPEIAE

This star lies at the center of Cassiopeia's W figure. Normally the constellation's third brightest star, it is an irregular variable. It is slowly losing mass into a disk or shell that surrounds it, and alterations in the shell's thickness might be responsible for its variations in brightness.

M 52   This group of a hundred or so stars is one of the richest

in the northern half of the sky, but only one of several open clusters scattered throughout Cassiopeia.

NGC 663   This small open cluster of quite faint stars is an attractive sight in a small telescope.

231

Centaurus

sen-TOR-us

The Centaur

These stars represent Chiron, one of the Centaurs—creatures that were half man, half horse. Unlike the other Centaurs, who were brutal, Chiron was extremely wise, and tutored Jason and Hercules.

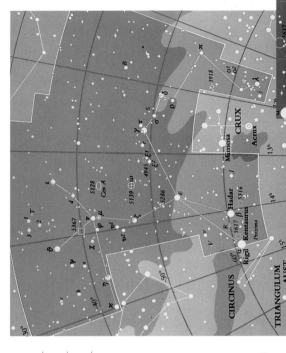

ALPHA (α) CENTAURI At the foot of the Centaur, this star is only 4.3 light-years away and is the Sun's nearest neighbor. One of the prettiest binary stars, its two components revolve around each other once every 80 years. The separation is around 20 arcseconds, but this will close to 2 arcseconds by about the year 2035. Alpha (α) and Beta (β) Centauri are the bright "pointers" to the Southern Cross.

PROXIMA CENTAURI A tiny red dwarf only 25,000 miles (40,000 km) across, this star is actually a little closer to us than the two stars of Alpha (α) Centauri, but it is thought to be their companion. It flares occasionally, jumping by half a magnitude or more, usually returning to its normal brightness within half an hour.

OMEGA (ω) CENTAURI This is perhaps the finest globular cluster in the entire sky. With the naked eye it is visible as a fuzzy star of magnitude 4. Unlike most globulars, it is oval rather than round in shape. A 3 inch (75 mm) telescope will show a large, fuzzy disk with mottled edges, while a 6 inch (150 mm) one will resolve it into stars. Viewed in an even larger telescope, under a dark sky, it looks magnificent, with the field of a low-power eyepiece overflowing with faint stars.

NGC 5128 Located only 4½ degrees north of Omega (ω) Centauri, this galaxy is distinguished by a dark band that runs across its center—probably the result of a collision with another galaxy. It is a strong source of radio energy, known to radio astronomers as Centaurus A. The dust lane is apparent in dark skies with a 4 inch (100 mm) or larger telescope.

NGC 3918 Not far from the Southern Cross, this planetary nebula presents a classic blue-green disk 12 arcminutes across, like a larger version of Uranus.

NGC 5128 A dark band of dust cuts across NGC 5128. This galaxy emits more than a thousand times the radio energy of our own Galaxy.

CENTAURI (CEN)
On meridian
10 p.m. May 10

233

Cepheus

SEE-fee-us

The King

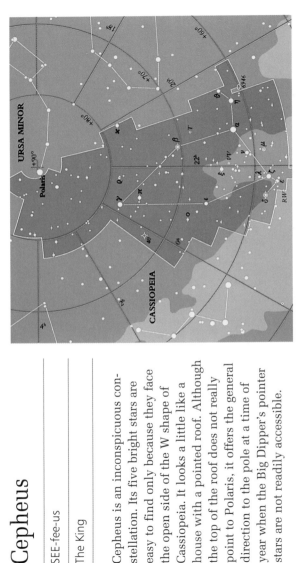

CEPHEI (CEP)
On meridian
10 p.m. Oct 1

Cepheus is an inconspicuous constellation. Its five bright stars are easy to find only because they face the open side of the W shape of Cassiopeia. It looks a little like a house with a pointed roof. Although the top of the roof does not really point to Polaris, it offers the general direction to the pole at a time of year when the Big Dipper's pointer stars are not readily accessible.

DELTA (δ) CEPHEI This star is the prototype for the Cepheid variables. Its highest magnitude is 3.5, as bright as neighboring Zeta (ζ) Cephei, and it fades to 4.4, the brightness of Epsilon (ε) Cephei. It completes a cycle every 5.4 days.

MU (μ) CEPHEI This star is so strikingly red that William Herschel called it the Garnet Star. Using Zeta (ζ) and Epsilon (ε) as comparison stars, you can watch it vary in brightness irregularly over hundreds of days.

234

Cetus

SEE-tus

The Whale; The Sea Monster

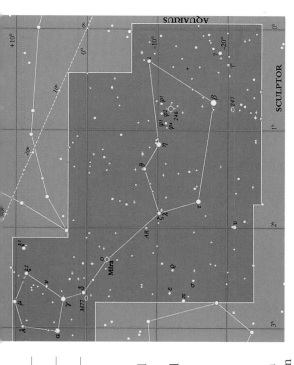

The constellation Cetus consists of faint stars, but it occupies a large area of sky. This sea monster's head is a group of stars not far from Taurus and Aries, and his body and tail lie toward Aquarius.

MIRA Omicron (o) Ceti, known as Mira, is the most famous long-period variable of all. On August 13, 1596, David Fabricius, a Dutch skywatcher, saw a new star in Cetus. Over the following weeks, it faded, disappeared, then reappeared in 1609. In 1662 Johannes Hevelius named it Mira Stella, the Wonderful Star. Mira varies from a magnitude of about 3.4 to a minimum of 9.3 over 11 months.

M 77 Cetus contains several galaxies, the brightest of which is M 77, near Delta (δ) Ceti. It is a 9th-magnitude spiral galaxy with a bright core. A 4 inch (100 mm) telescope shows a faint circular disk around the core.

235

Chamaeleon

ka-MEE-lee-un

The Chameleon

Johann Bayer drew this constellation early in the 17th century, following descriptions given by early south sea explorers. One of the smallest and least conspicuous of the constellations, Chamaeleon does a good job hiding in the sky. Consisting of a few faint stars, it lies close to the south celestial pole, south of Carina and right beside the south polar constellation of Octans.

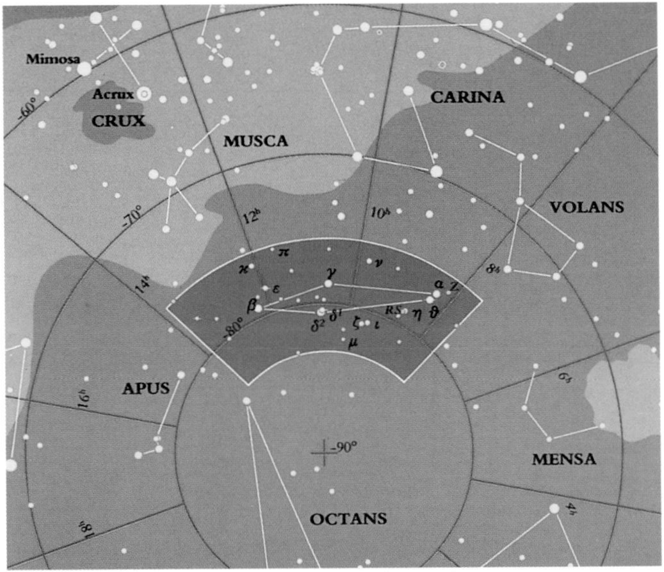

CHAMAELEONTIS
(CHA)
On meridian
10 p.m. April 1

Z CHAMAELEONTIS This faint variable star erupts periodically. At its minimum it shines at magnitude 16.2, invisible except in a telescope with an aperture of at least 12 inches (300 mm). However, every three to four months it under-goes an outburst, rising within just a few hours to about magnitude 11.5, and for a few days it is visible through a 6 inch (150 mm) telescope. Even so, this does not constitute an easy target in a corner of the sky with few stars.

Circinus

SUR-seh-nus

The Drawing Compass

The early explorers in the south seas were less interested in mythology than in the modern instruments on which they relied to find their way around uncharted waters. The Drawing Compass is one of a number of obscure constellations that were designated by the French astronomer Nicolas-Louis de Lacaille. He worked at an observatory at the Cape of Good Hope from 1750 to 1754, where he compiled a catalog of more than 10,000 stars. The constellation is not visible to most northern observers.

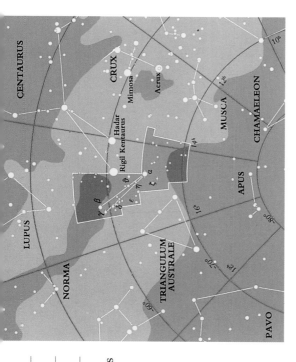

ALPHA (α) CIRCINI This, at only 3rd magnitude, is the constellation's brightest star and chief attraction. It lies just near the much brighter Alpha (α) Centauri. It is about 65 light-years away and has a faint 9th-magnitude companion. Alpha (α) Circini represents the hinge of the compass; Beta (β) Circini is one of the points.

CIRCINI (CIR)
On meridian
10 p.m. June 1

Columba

koh-LUM-bah

The Dove

Immediately south of Canis Major, the constellation Columba was named by Petrus Plancius, a 16th-century Dutch theologian and map-maker. This inconspicuous group of stars honors the dove that Noah sent out from the ark after the rains had stopped, to see if it could find dry land. According to at least one source, the name for the constella-tion's alpha (α) star, Phakt, is the Arabic word for "ring dove."

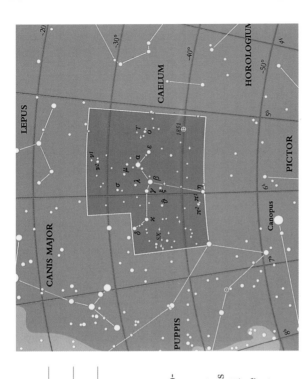

COLUMBAE (COL)
On meridian
10 p.m. Jan 20

NGC 1851 Bright and large, this 7th-magnitude globular cluster appears as a misty spot through binoculars under a good sky. A 6 inch (150 mm) telescope will begin to resolve the cluster's brightest stars.

T COLUMBAE This star is a Mira variable, with a maximum magnitude of 6.7. It drops down to magnitude 12.6 and then rises again over a period of seven and a half months.

Coma Berenices

KOH-mah bear-eh-NEE-seez

Berenice's Hair

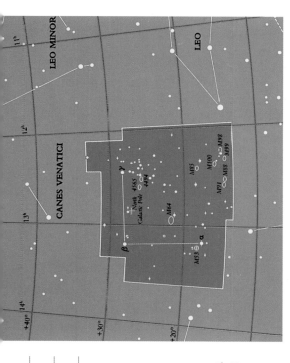

Coma Berenices has no bright stars and is hard to distinguish, but it is a remarkable area of sky. It is a sprinkling of faint stars super-imposed on a cloud of galaxies.

M 53   This fine globular cluster is about 3 arcminutes in diameter and lies close to Alpha (α) Comae Berenices.

THE BLACKEYE GALAXY (M 64)   This looks like an ordinary spiral galaxy, with tightly wound arms, but seen with a 4 to 6 inch (100 to 150 mm) telescope or larger, a huge cloud of dust can be seen dominating its center, giving it the look of a black eye.

NGC 4565   Under a dark sky, a small telescope should show this 10th-magnitude object as a pencil-thin line of haze. It is a spiral galaxy seen edge-on, with a dust lane that becomes apparent in 8 inch (200 mm) telescopes.

Corona Australis

kor-OH-nah os-TRAH-lis

The Southern Crown

This small semicircular group of faint stars lies just south of Sagittarius and is said to represent a crown of laurel or olive leaves. One story relating to the crown comes from Ovid's *Metamorphoses*. Juno discovered that her husband, Jupiter, was the lover of Semele, a human. Masquerading as Semele's maid, Juno suggested that Semele ask Jupiter to appear before her in all his glory. Jupiter was appalled at her request, but did not refuse it. When she saw him in his splendor she was consumed by fire. Her unborn child was saved, however, to become Bacchus, god of wine, who honored his mother by placing the crown in the sky.

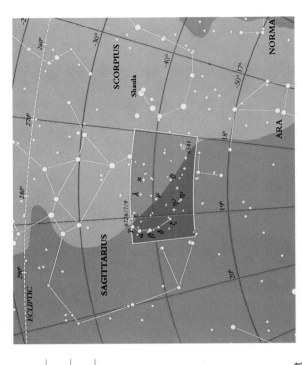

CORONAE AUSTRALIS (CrI)
On meridian
10 p.m. Aug 1

NGC 6541 Viewed through smaller telescopes, this globular cluster looks like a small nebulous disk. An 8 inch (200 mm) telescope only begins to resolve the edge into stars. The cluster is located about 22,000 light-years away.

Corona Borealis

kor-OH-nah bor-ee-AL-is

The Northern Crown

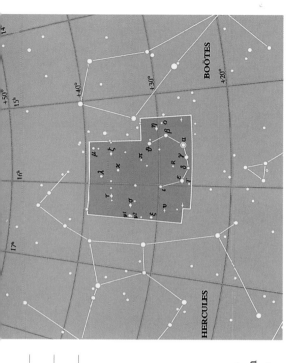

Just 20 degrees northeast of Arcturus lies the Northern Crown, a small semicircle of stars that are faint but very distinct. In Greek mythology, the crown belongs to Ariadne, daughter of Minos, King of Crete.

R CORONAE BOREALIS One of the more remarkable stars in the sky, R Cor Bor, as it is generally known, is a nova in reverse. Normally shining at magnitude 5.9, at completely irregular intervals the star will suddenly fade, sometimes by as much as eight magnitudes, as dark material erupts in its atmosphere. It then slowly recovers as the material dissipates.

T CORONAE BOREALIS Now shining at magnitude 10.2, in 1866 this star suddenly rose to magnitude 2. Known as a recurrent nova, the star repeated the performance unexpectedly in 1946, and will probably do so again.

241

Corvus and Crater

KOR-vus   KRAY-ter

The Crow   The Cup

Arc to Arcturus, speed to Spica, then turn west and you will see a small foursome of stars that the ancients called the Crow or the Raven. Crater is a fainter constellation alongside that looks like a cup, and that represents the chalice of the Greek god Apollo.

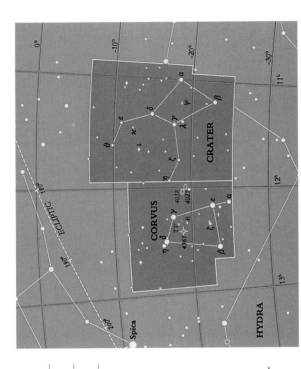

CORVI (CRV)
CRATERIS (CRT)
On meridian
10 p.m. April 20

R CORVI This Mira-type variable star ranges from magnitude 6.7 to 14.4 over a period of about 10 months.

THE RING-TAILED GALAXY Also called the Antennae or Rat-tailed Galaxy, NGC 4038 and NGC 4039 form a faint, 11th-magnitude pair of galaxies that are in the process of interacting or colliding. Needing an 8 inch (200 mm) telescope to see, it is still one of the brightest pairs of connected galaxies. Long-exposure photographs show the galaxies have a pair of "tails," hence the common names.

Crux

KRUKS

The Southern Cross

The most famous southern constellation, the Southern Cross appears on the flags of several nations.

ACRUX Acrux, or Alpha (α) Crucis, is the double star at the foot of the cross, separated by 4½ arcseconds. A third star, of 5th magnitude, lies 90 arcseconds away.

GAMMA (γ) CRUCIS This optical double consists of a magnitude 6.4 star lying almost 2 arcminutes from an orange primary.

THE JEWEL BOX Superimposed on Kappa (κ) Crucis, this is one of the finest open clusters. Although small, it sparkles in any instrument, and has several stars of contrasting color.

THE COAL SACK This large and dense dark nebula is clearly visible in a dark sky against the star clouds of the Milky Way.

CRUCIS (CRU)
On meridian
10 p.m. May 1

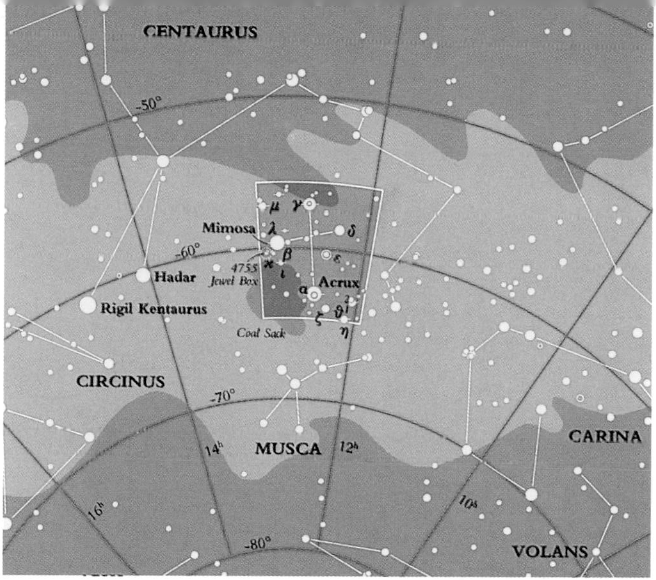

Cygnus

SIG-nus

The Swan

Cygnus is the Northern Hemisphere's answer to Crux. Looking like a large cross, Cygnus straddles the northern Milky Way.

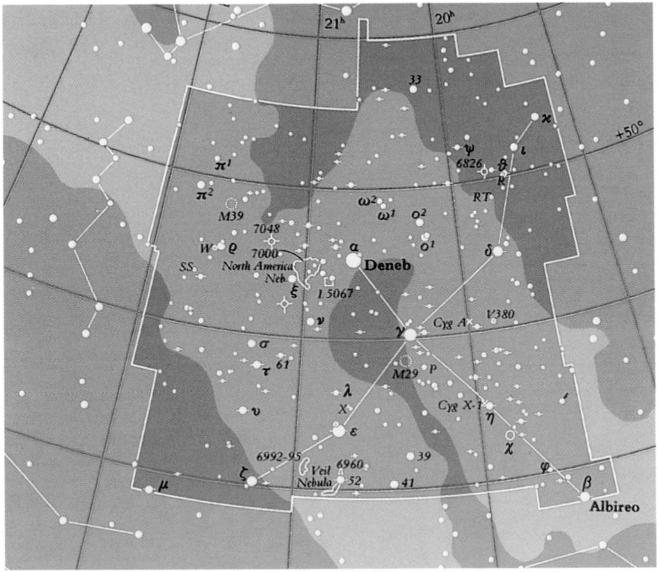

◈ DENEB (ALPHA [α] CYGNI) In Arabic Deneb means "tail," which is where this star is positioned on the swan. This mighty star is 25 times more massive and 60,000 times more luminous than the Sun.

◈ ALBIREO (BETA [β] CYGNI) The unaided eye sees Albireo as one star; a telescope transforms it into a double with a separation of 34 arcseconds. One member is yellow with a magnitude of 3, and the other is bluish with a magnitude of 5.

CYGNI (CYG)
On meridian
10 p.m. Aug 20

✂ 61 CYGNI Dubbed the Flying Star because of its rapid motion relative to more distant stars, this double is easily separated in small telescopes. The two components revolve around each other over the course of about 650 years.

THE NORTH AMERICA NEBULA (NGC 7000) This giant cloud is illuminated by Deneb, which lies only 3 degrees to the west. Because of its size, the nebula is difficult to see in a telescope: it is best seen with the naked eye on a dark night. Photographs show this nebula to look surprisingly like the shape of North America, but this resemblance is not readily apparent to the eye when observing.

M 39 This open star cluster is seen at its best through binoculars. On a clear night you might be able to see it with the naked eye.

CIII (χ) CYGNI At maximum brightness, magnitude 4 or 5, this long-period variable is bright enough to be seen with the naked eye. It fades to about magnitude 13 and then climbs back in a period of a little more than 13 months.

THE VEIL NEBULA (NGC 6960, 6992, 6995) The lacy remnants of an ancient supernova, this beautiful nebulosity requires at least a 6 inch (150 mm) telescope. NGC 6960, the nebula's western arc, passes through the star 52 Cygni, which makes it easier to find but harder to see. The Veil Nebula lies 2,500 light-years away.

VEIL NEBULA Also known as the Cygnus Loop, the Veil Nebula is the expanding blast-wave from a supernova that exploded about 15,000 years ago.

Delphinus

del-FIE-nus

The Dolphin

This small group of faint stars has a distinctive shape, a little like a kite. Its alpha (α) star is named Sualocin and its beta (β) star is known as Rotanev. These names honor a relatively recent observer, Niccolo Cacciatore, long-time associate of the famous 19th-century observer Giuseppe Piazzi. Star atlases at the time included these names without comment, but the Reverend Thomas Webb worked out that the names, spelled backward, are Nicolaus Venator—the Latinized version of Cacciatore's name.

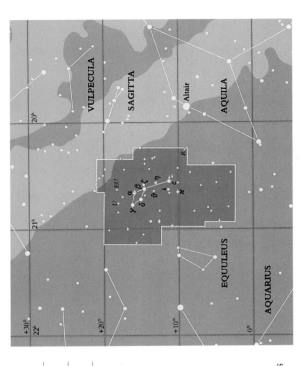

DELPHINI (DEL)
On meridian
10 p.m. Sept 1

brighter component is magnitude 4.5 and the fainter, which is slightly green in color, is 5.5.

GAMMA (γ) DELPHINI This is an optical double with a separation of 10 arcseconds. The

R DELPHINI This Mira star has a magnitude range of 8.3 to 13.3 over a period of 285 days.

Dorado

doh-RAH-doh

The Goldfish; The Dolphinfish

Lying far to the south, this constellation was first recorded by Bayer in his star atlas of 1603.

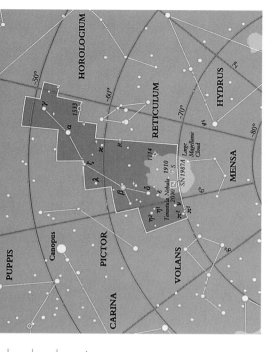

DORADUS (DOR)
On meridian
10 p.m. Jan 1

◈ THE LARGE MAGELLANIC CLOUD (LMC) This is a companion galaxy to the Milky Way, lying only 163,000 light-years away. The LMC spans about 11 degrees of the sky. It was from this galaxy that supernova 1987A blazed forth. The LMC is plainly visible in a dark sky, but it is easily lost in the glare of city lights.

✺ THE TARANTULA NEBULA (NGC 2070) This is one of the finest emission nebulas in the sky. It is enormous, perhaps 30 times the size of the more famous Orion Nebula (M 42).

✺ S DORADUS Located within the open cluster NCC 1010, this star varies irregularly in brightness between magnitudes 8 and 11.

Draco

DRAY-koh

The Dragon

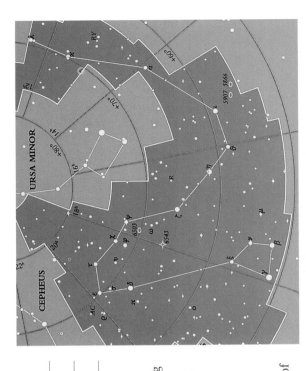

This constellation is circumpolar from much of the Northern Hemisphere and is best seen during the warmer months of the year. Large and faint, the Dragon is hard to trace as it winds about between the constellations of Ursa Major, Boötes, Hercules, Lyra, Cygnus, and Cepheus.

QUADRANTIDS This is one of the strongest meteor showers. The time of maximum activity is around January 3, although the shower lasts only a few hours.

DRACONIDS The Draconids meteor shower reaches its peak around October 8.

DRACONIS (DRA)
On meridian
10 p.m. July 1

NGC 6543 This 8th-magnitude planetary nebula lies midway between the stars Delta (δ) and Zeta (ζ) Draconis. It is bright blue-green in color, but high power is needed in order to make out its small, hazy disk.

Equuleus

eh-KWOO-lee-us

The Little Horse

With the exception of Crux, Equuleus occupies a smaller patch of sky than any other constellation. It lies just to the southeast of Delphinus and because it has no bright stars it is of limited interest. The famous Greek astronomer Hipparchus is thought to have made up the constellation in the second century BC. It has been said to represent Celeris, brother of Pegasus (the Winged Horse), given to Castor (one of the twins represented by Gemini) by Mercury.

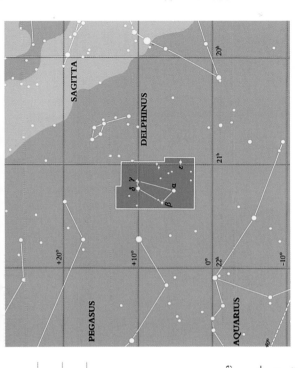

ALPHA (α) EQUULEI The constellation's brightest star is named Kitalpha, which is Arabic for "little horse."

EQUULEI (EQU)
On meridian
10 p.m. Sept 1

249

Eridanus

eh-RID-an-nus

The River

This constellation has been seen as a river since ancient times—usually the Euphrates or the Nile. In *Metamorphoses*, the Roman poet Ovid writes of Phaethon being tossed out of the chariot of the Sun to drown in Eridanus.

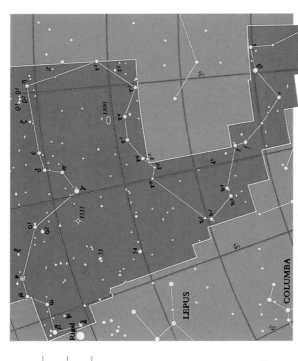

ERIDANI (ERI)
On meridian
10 p.m. Dec 10

⟡ EPSILON (ε) ERIDANI Only 10.8 light-years away, this is one the closest stars to Earth. It is a smaller version of our Sun. Radio telescopes have searched for but been unsuccessful in discovering signals that indicate intelligent life.

✦ OMICRON 2 (o₂) ERIDANI This remarkable triple consists of a 4th-magnitude orange dwarf, a 9th-magnitude white dwarf, and an 11th-magnitude red dwarf. The red and white dwarfs form a pair (separation 8 arcseconds), and are separated from the brighter star by 80 arcseconds. The white dwarf is the only one of its class that is easy to see in a small telescope.

Fornax

FOR-nax

The Furnace

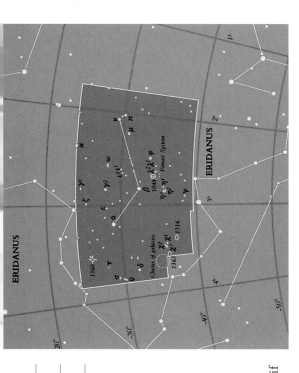

FORNACIS (FOR)
On meridian
10 p.m. Dec 1

This constellation was originally named Fornax Chemica, the Chemical Furnace, in honor of the chemist Antoine Lavoisier, who was guillotined during the French Revolution in 1794. Today it is simply known as the Furnace.

THE FORNAX GALAXY
CLUSTER   While there are no bright points of interest in Fornax, if you have a large telescope you will enjoy this challenging cluster of galaxies near the Fornax–Eridanus border. With a wide-field eyepiece, you may see up to nine galaxies in a single field of view. The brightest galaxy, at 9th magnitude, NGC 1316, is also the radio source Fornax A.

THE FORNAX SYSTEM   This dwarf galaxy is too faint to see with a backyard telescope, but one of its globular clusters, NGC 1049, is, at magnitude 12.9, visible in a 10 inch (250 mm) telescope under a good sky.

251

Gemini

JEM-eh-nye

The Twins

The Greeks named Gemini's two brightest stars Castor and Pollux after the twins who were the offspring of Leda, princess of Sparta, and Zeus, king of the gods.

CASTOR (ALPHA [α] GEMINORUM) This sextuple star can be seen only as a double through a small telescope.

ETA (η) GEMINORUM This bright semi-regular variable ranges from magnitude 3.2 to 3.9 and back over about eight months.

M 35 This open cluster is beautiful through binoculars and spectacular in a small telescope.

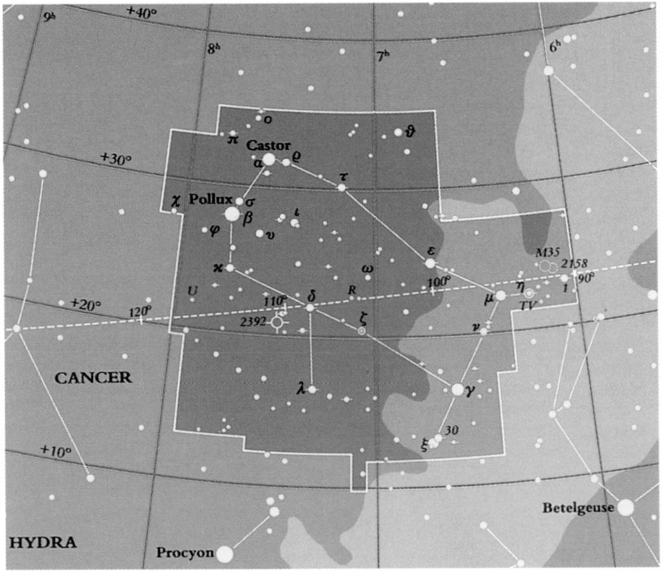

NGC 2158 is a smaller, fainter open cluster on its southwest edge.

THE CLOWNFACE OR ESKIMO NEBULA (NGC 2392) This 8th-magnitude planetary nebula has a bright central star.

GEMINORUM (GEM)
On meridian
10 p.m. Feb 1

Grus

GROOS

The Crane

Grus has little to offer the small telescope user, although some faint galaxies provide targets for telescopes of 8 inch (200 mm) aperture or larger. Grus has only three fairly bright stars, which can be used as a simple illustration of magnitude.

ALPHA (α) GRUIS Also known as Alnair, this is a large, blue main-sequence star over a hundred times as luminous as the Sun. Being 100 light-years away, it is the brightest of the three stars only because it is relatively close to us.

BETA (β) GRUIS This is a much larger red giant star, several hundred times as luminous as the Sun, but

its orange red color makes it appear fainter than Alpha (α) Gruis.

GAMMA (γ) GRUIS This blue giant is more luminous, than the others, and appears fainter than them, as it is 200 light-years away.

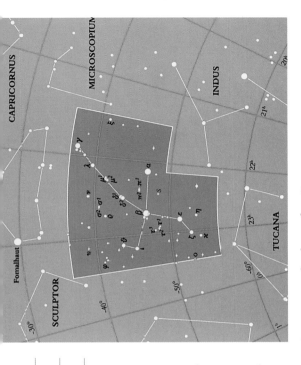

253

Hercules

HER-kyu-leez

Hercules

For northern observers, Hercules, with its "keystone" of four stars—Epsilon (ε), Zeta (ζ), Eta (η) and Pi (π)—is one of the best of the summer constellations.

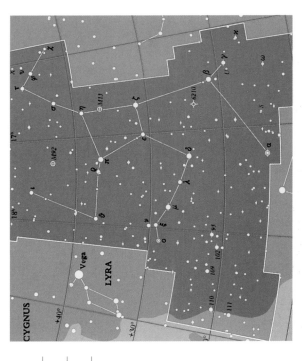

◈ THE HERCULES CLUSTER (M 13) This globular cluster is faintly visible to the naked eye as a fuzzy spot, but through a telescope it is a sight to behold. The edges begin to resolve into stars in a 6 inch (150 mm) telescope. When you view this cluster, you are looking 23,000 years into the past.

◈ M 92 M 13's slightly smaller and fainter cousin, this cluster of stars is 26,000 light years away.

◈ RAS ALGETHI (ALPHA [α] HERCULIS) This is a very red star, varying from magnitude 3.1 to 3.9. It is also a splendid colored double, with a 5th-magnitude blue-green companion about 5 arcseconds away from an orange primary.

Horologium

hor-oh-LOH-jee-um

The Clock

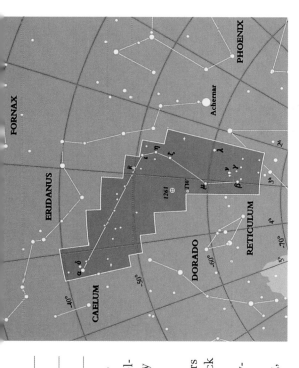

HOROLOGII (HOR)
On meridian
10 p.m. Dec 10

A small group of stars lying east of Archernar, this is one of the constellations mapped by the 18th-century French astronomer Nicolas-Louis de Lacaille. Originally called Horologium Oscillatorium, it honors the invention of the pendulum clock by the Dutch scientist Christiaan Huygens in 1656 or 1657. By applying the law of the pendulum, discovered by Galileo, to clockmaking, he significantly increased the accuracy of timekeeping.

R HOROLOGII This long-period variable star was discovered from an observing station that Harvard University used to run in Peru. In 13½ months, it completes its cycle of variation from 5th to 14th magnitude and back.

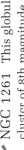 NGC 1261 This globular cluster of 8th magnitude is only 6 arcminutes across, so is a target for a larger telescope.

255

Hydra

HY-dra

The Sea Serpent

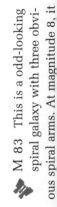

HYDRA Although in mythology, Hydra was the nine-headed serpent killed by Hercules, it is generally portrayed as a sea serpent. It snakes a quarter of the way across the sky.

In Greek mythology, Hydra was the nine-headed serpent that Hercules had to kill as one of his 12 labors. Each time he lopped off one head, two others grew in its place. Hercules emerged from this nightmare by having his nephew burn the stump of each severed neck, preventing new heads from sprouting.

 **R HYDRAE** This Mira star's light changes were first seen in the late 1600s. It varies over 13 months from a maximum of magnitude 3.5 to a minimum of 10.9.

V HYDRAE A rare example of a carbon star, this is a low-temperature red giant producing carbon. It is so deeply red that you can be sure you have found it merely by its color. The star varies erratically between magnitudes 6 and 12, with two superimposed periods—one about 18 months, and the other 18 years.

M 48 (NGC 2548) Long thought to be a missing Messier object because he wrongly reported its position, M 48 is now considered to be the same as NGC 2548—a large open cluster best seen using binoculars or a wide-field telescope.

M 83 This is a odd-looking spiral galaxy with three obvious spiral arms. At magnitude 8, it

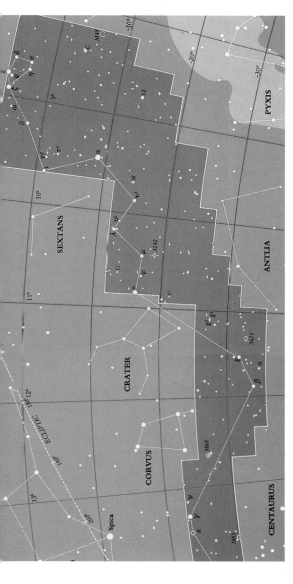

✂ THE GHOST OF JUPITER NEBULA (NGC 3242) This nebula is the brightest planetary nebula in this part of the sky. It is about 16 arcseconds across and shows its structure well in a 10 inch (250 mm) telescope.

is one of the brighter galaxies visible in binoculars and will show more detail at higher magnification in a telescope. Keep a lookout for "new" stars here, as M 83 has been the location of four supernovas in the last 60 years.

Hydrus

HY-drus

The Water Snake

The German astronomer Johann Bayer created this constellation, publishing it in his 1603 star atlas. He placed it near Achernar, the mouth of the River Eridanus, between the Large and Small Magellanic Clouds. It is sometimes called the Male Water Snake, to avoid confusion with Hydra.

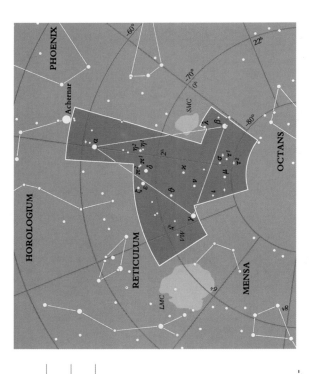

HYDRI (HYI)
On meridian
10 p.m. Dec 1

VW HYDRI This star is the most popular cataclysmic variable with Southern Hemisphere observers. When in its usual state, it shines at a faint 13th magnitude, but when it goes into outburst, an event which occurs about once a month, it can become brighter than 8th magnitude in just a few hours.

Indus

IN-dus

The Indian

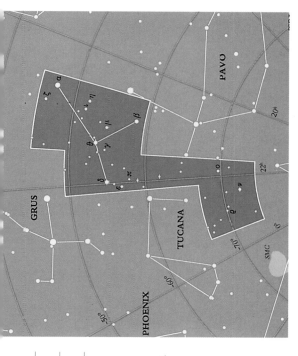

This constellation was added to the southern sky by the German astronomer Johann Bayer to honor the Native Americans that European explorers encountered on their travels. The figure of Indus is positioned between three birds: Grus, the Crane; Tucana, the Toucan; and Pavo, the Peacock.

EPSILON (ε) INDI   Only 11.3 light-years away, this is one of the closest stars to the Sun and is somewhat similar to it. With four-fifths of the Sun's diameter and one-eighth its luminosity, scientists consider Epsilon (ε) Indi to be worth investigating for planets and for evidence of extraterrestrial intelligence, such as radio signals. In the early 1960s, when Frank Drake began searching for signs of life elsewhere in the galaxy, he used this star as one of his targets. In 1972, the Copernicus Satellite searched unsuccessfully for laser signals from this star.

259

Lacerta

lah-SIR-tah

The Lizard

Lacerta is far enough north to be circumpolar at the higher mid-northern latitudes. It lies, inconspicuously, to the south of Cepheus. The German astronomer Johannes Hevelius suggested that this group of stars be named Lacerta, the Lizard, in his star catalog published in 1690.

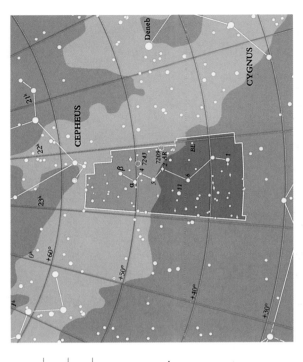

BL LACERTAE   Since this object varies from 13.0 to 16.1, it is invisible to any but the largest amateur telescopes. However, BL Lacertae is worth a look, since it is not a star at all but the nucleus of a distant elliptical galaxy. Some of this class of BL Lacertae-type (BL Lac) objects have been known to change by as much as two magnitudes in a single day. Theories suggest that BL Lac objects, quasars, and other high-powered galaxies are all closely related "active galaxies." The powerful energy source at the center may be a black hole.

Leo and Leo Minor

LEE-oh   LEE-oh MY-ner

The Lion   The Little Lion

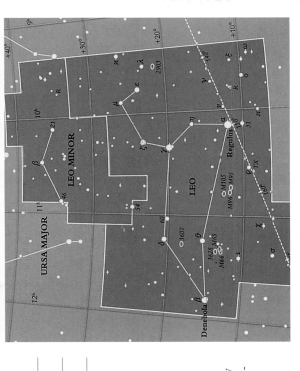

Unlike many constellations, Leo, with its sickle tracing out a head, really can be pictured as its namesake, a lion. Leo Minor was introduced during the 17th century.

🔭 GAMMA (γ) LEONIS This double star has orange-yellow components of 2nd and 3rd magnitude separated by 5 arcseconds.

👓 R LEONIS A Mira variable, this ranges from magnitude 5.9 to 11 over about 10½ months.

👓 R LEONIS MINORIS Another Mira star, this one takes about a year to vary between magnitudes 7.1 and 12.6.

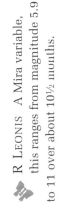

LEONIS (LEO)
LEONIS MINORIS (LMI)
On meridian
10 p.m. April 1

👓 M 65 AND M 66 These two spiral galaxies near Theta (ϑ) Leonis are visible in binoculars but give a better view in a telescope.

👁 LEONIDS This meteor shower peaks annually on November 17.

Lepus

LEE-pus

The Hare

A faint constellation, Lepus is nevertheless easy to find because it is directly south of Orion.

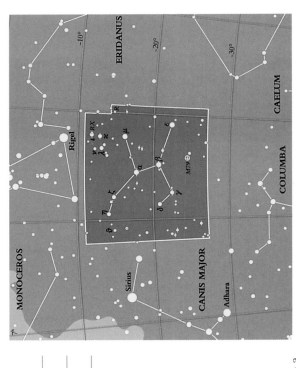

GAMMA (γ) LEPORIS Easy to separate in virtually any telescope, this wide double star with contrasting colors has a separation of 96 arcseconds.

HIND'S CRIMSON STAR Likened by some observers to a drop of blood in the sky, R Leporis is the variable that the 19th-century British astronomer J. Russell Hind called the Crimson Star. Over a period of 14 months, the star varies in magnitude from a maximum of as much as 5.5 to a minimum of 11.7. Its coloring is at its most striking when the sky is dark and the star is near maximum brightness.

M 79 This globular cluster is best observed using a telescope 8 inches (200 mm) or larger.

Libra

LEE-bra

The Scales; The Balance

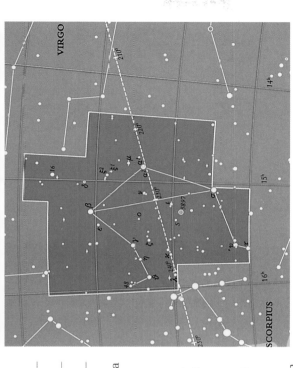

Looking like a high-flying kite, Libra is easy to find by extending a line westward from Antares and its two bright neighbors in Scorpius. The line reaches a point between Alpha (α) and Beta (β) Librae. Libra is one of the constellations of the zodiac and was associated with Themis, the Greek goddess of justice, whose symbol was a pair of scales.

Originally these stars were seen as the claws of Scorpius. But Roman skylore fancied placing a set of scales here, reviving a figure that dated from Sumerian times.

LIBRAE (LIB)
On meridian
10 p.m. June 10

2.3 days, from 4.9 to 5.9. The entire cycle is visible to the naked eye.

S LIBRAE A Mira star, S Librae varies from an 8.4 maximum to a 12.0 minimum over a period of just over six months.

DELTA (δ) LIBRAE Similar to Algol, this eclipsing variable star fades by about a magnitude every

263

Lupus

LOO-pus

The Wolf

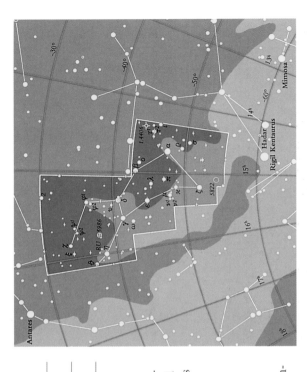

South of Libra and east of Centaurus, Lupus, the Wolf, is a small constellation with some 2nd-magnitude stars. It is almost joined with Centaurus, as if the Centaur is stroking the wolf like a pet. The ancient Greeks and Romans called this group of stars Therion—an unspecified wild animal. Lying within the band of the Milky Way, this constellation is home to a number of open and globular clusters.

RU LUPI (Read this name out loud. After immersing yourself in constellation lore to this point, you possibly are!) RU Lupi is a faint nebular variable, with a maximum of only 9th magnitude. Its irregular variation is characteristic of young stars still wrapped in nebulosity.

 NGC 5986 This globular cluster is visible in binoculars, and sits close to some 6th- and 7th-magnitude stars.

LUPI (LUP)
On meridian
10 p.m. June 10

Lynx

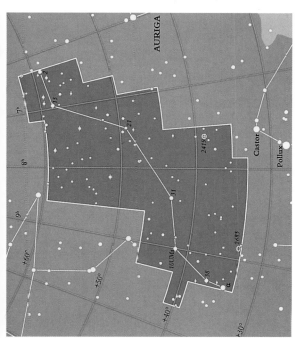

LYNCIS (LYN)
On meridian
10 p.m. Feb 20

With only one 3rd-magnitude star, Lynx is one of the hardest constellations to find. Johannes Hevelius charted this figure around 1690, apparently naming it Lynx because you need to have the eyes of a lynx to spot it. The same is true of its deep-sky objects.

THE INTERGALACTIC TRAMP (NGC 2419) Lying some 7 degrees north of Castor, the brightest star in Gemini, this is a very faint and distant globular cluster. It is more than 60 degrees from any other globular. At 210,000 light-years, it is more distant than the Large Magellanic Cloud and is so far away that it might escape the gravitational pull of our galaxy. It is for this reason that the U.S. astronomer Harlow Shapley called it the Intergalactic Tramp. Through a 10 inch (250 mm) or larger telescope, NGC 2419 appears as a fuzzy knot of light.

Lyra

LYE-rah

The Lyre

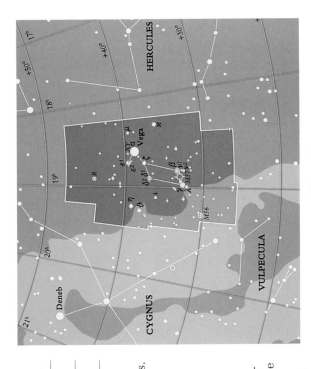

This constellation is dominated by Vega, one of the sky's brightest stars. You can imagine the lyre strings stretched across the parallelogram of four stars that accompany it.

EPSILON (ε) LYRAE This is a "double double" star. The slightest optical aid shows two 5th-magnitude stars—ε¹ and ε². Both are themselves doubles, with separations under 3 arcseconds. A 4 inch (100 mm) telescope operating at a magnification of 100 or more will split both of them.

BETA (β) LYRAE This eclipsing variable ranges from magnitude 3.3 to 4.4 in 13 days.

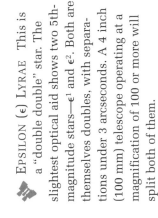

THE RING NEBULA (M 57) This planetary nebula lies between Beta (β) and Gamma (γ) Lyrae. It appears in 3 inch (75 mm) telescopes as a star out of focus at low magnification. Higher power will show its ring shape.

LYRAE (LYR)
On meridian
10 p.m. Aug 1

Mensa

MEN-sah

The Table; The Table Mountain

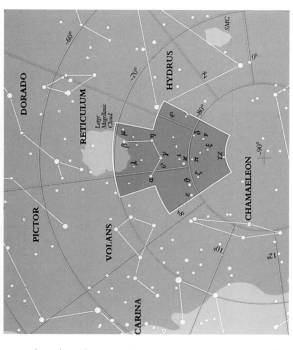

The only constellation that refers to a specific piece of real estate, Mensa was originally called Mons Mensae by Nicolas-Louis de Lacaille, after Table Mountain, south of Cape Town, South Africa, where he did a good deal of his work. He created this small constellation from stars between the Large Magellanic Cloud and Octans. The northernmost stars of the constellation, representing the summit of the mountain, are hidden in the Large Magellanic Cloud, in the same way that Table Mountain is often shrouded in clouds.

close to us, its light taking only 28 years to reach Earth.

◈ BETA (β) MENSAE Lying very near the Large Magellanic Cloud, this faint star of magnitude 5.3 is 155 light-years away.

◈ ALPHA (α) MENSAE This dwarf star has an apparent magnitude of 5.1. It lies comparatively

Microscopium

my-kro-SKO-pee-um

The Microscope

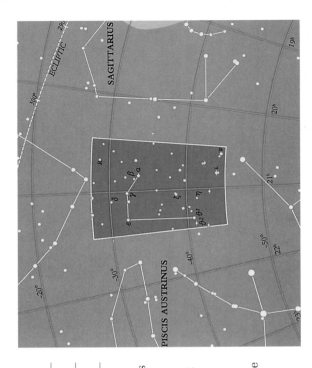

This small, faint constellation, which lies just south of Capricornus and east of Sagittarius, is made up of 5th-magnitude stars. A modern creation, it was formulated in about 1750 by the French astronomer Nicolas-Louis de Lacaille. It commemorates the microscope, the invention of which is credited to the Dutch spectacle-maker Zacharias Janssen, around 1590, and to Galileo, among others.

R MICROSCOPII This faint Mira variable star has a rapid cycle lasting only 4½ months, during which it drops from magnitude 9.2 to 13.4 and then climbs back again.

Monoceros

moh-NO-ser-us

The Unicorn

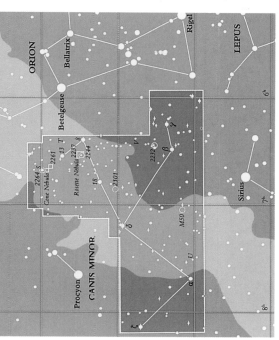

MONOCEROTIS
(MON)
On meridian
10 p.m. Feb 1

This faint constellation was formed in about 1624 by the German Jakob Bartsch. Monoceros is the Latin form of a Greek word meaning "one-horned," a reference to the unicorn.

M 50 This beautiful open cluster, lying slightly more than one-third of the way from Sirius to Procyon, is easy to find. Some of the cluster's stars are arranged in pretty arcs.

THE ROSETTE NEBULA (NGC 2237) In a 10 inch (250 mm) telescope, this ring-shaped nebula, and the open cluster it contains (NGC 2244), offer a scene of delicate beauty. Binoculars and small telescopes will reveal the nebula on very clear nights.

 THE CHRISTMAS TREE CLUSTER (NGC 2264) This open cluster really does look like a Christmas tree.

269

Musca

MUSS-kah

The Fly

Musca is an easy constellation to find, just to the south of the Southern Cross.

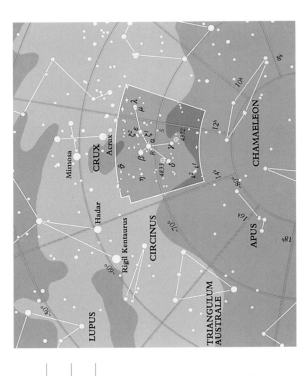

BETA (β) MUSCAE This elegant double star consists of two 4th magnitude stars that revolve around each other in a period that spans several hundred years. The pair is some 520 light-years from Earth. The separation of 1.6 arcseconds is very tight, presenting a challenge for a 4 inch (100 mm) telescope.

NGC 4372 This globular cluster is close to Gamma (γ) Muscae and has faint stars spread over 18 arcminutes.

NGC 4833 This is a large, faint globular cluster, 18,000 light-years away. It lies within 1 degree of Delta (δ) Muscae. A 4 inch (100 mm) or larger telescope is needed to begin to resolve the cluster into individual stars.

Norma

NOR-muh

The Square

East of Centaurus and Lupus is a small constellation called Norma, the Square. When he named this group of stars, Nicolas-Louis de Lacaille decided to call it Norma et Regula, the Level and Square, after a carpenter's tools. Since those days, however, the Regula has been forgotten. The constellation lies alongside Circinus, the Drawing Compass, which he named at the same time. Set in the southern Milky Way, Norma presents good fields for binoculars, containing a number of open clusters. For such a small constellation, Norma has also been quite lucky with the appearance of novas: there was one in 1893 and another in 1920.

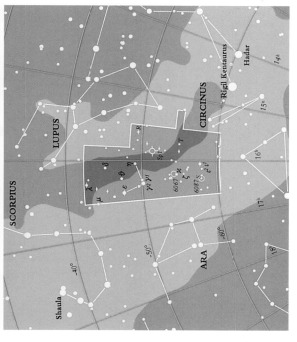

 NGC 6067 This is a small open cluster. Large binoculars or a telescope reveal about a hundred stars within a stunning field.

 NGC 6087 This is another of Norma's striking open clusters.

NORMAE (NOR)
On meridian
10 p.m. June 10

Octans

OCK-tanz

The Octant

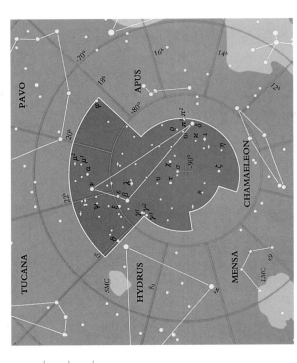

OCTANTIS (OCT)
On meridian
10 p.m. Sept 10

To honor the invention of the octant in 1730, Nicolas-Louis de Lacaille formed this south polar constellation. The forerunner of the sextant, the octant was used for measuring the altitude of a celestial body.

● SIGMA (σ) OCTANTIS This is the south pole star. At magnitude 5.4, it is barely visible to the naked eye on a dark night, so while it does mark the pole, it is not as convenient a marker star as the north's Polaris. The celestial poles move as the axis of the Earth precesses, or wobbles like a top, over some 26,000 years. Sigma (σ) Octantis was at its closest to the pole in about 1870, at just under half a degree. It is now just over 1 degree. In about another 3,000 years, the pole will begin to move through Carina, and it will pass near Delta (δ) Carinae in about 7,000 years. At 2nd magnitude, this is the brightest south pole star the Earth ever sees.

Ophiuchus

oh-fee-YOO-cuss

The Serpent Bearer

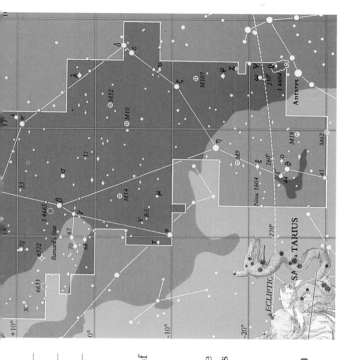

Ophiuchus, entwined with the constellation Serpens, covers a large expanse of sky and contains some of the Milky Way's richest star clouds.

M 9, 10, 12, 14, 19, 62
These globular clusters provide examples of different concentrations of stars. M 9 and 14 are rich; M 10 and 12 are looser; M 19 is oval; M 62 is somewhat irregular in outline. All are visible in binoculars but require a 6 or 8 inch (150 or 200 mm) telescope to do them justice.

RS OPHIUCHI This nova has had a number of outbursts. Its magnitude ranges from 11.8 to as high as 4.3 during outburts.

BARNARD'S STAR This magnitude 9.5 red dwarf star has the greatest proper motion of any known star.

Orion

oh-RYE-un

The Hunter

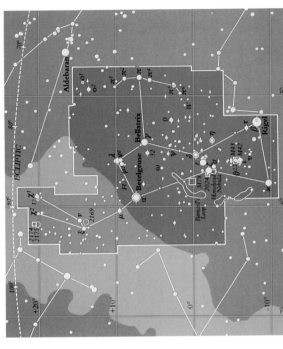

Orion is a treasure, with Rigel, Betelgeuse and its three belt stars in a row lighting up the sky from December to April.

◆ BETELGEUSE (ALPHA [α] ORIONIS) Betelgeuse (pronounced BET-el-jooze but sometimes corrupted to BEETLE-juice) is fabulous. (Its name comes from the Arabic for "hand of al-jauza," an obscure term which may refer to a female character in Arabic mythology.) A variable star, it ranges in magnitude from 0.3 to 1.2 over a period of almost seven years. However, the semi-regular nature of the variation means that it is often possible to detect changes over just a few weeks.

◆ RIGEL (BETA [β] ORIONIS) The name Rigel is derived from the Arabic word for "foot," a reference to the star's position in Orion. This mighty supergiant, some 800 light-years from Earth, is about 43,000 times as luminous as the Sun.

THE ORION NEBULA (M 42)

This star nursery, one of the marvels of the night sky, is also known as the Great Nebula. Plainly visible to the naked eye under a dark sky, it can be clearly seen through binoculars even in the city. The swirls of nebulosity spread out from its core of four stars called the Trapezium, which power the nebula. Photographs usually "burn out" the inner region of the nebula and obscure the Trapezium stars.

M 43 This is a small patch of nebulosity just north of the main body of the Orion Nebula. In fact, the M 42 complex is simply the brightest part of a gas cloud covering the constellation of Orion at a distance of some 1,500 light-years.

THE HORSEHEAD NEBULA (IC 434)

Also known as Barnard 33, this dark nebula is projected against a background of diffuse nebulosity, alongside the bright belt star Zeta (ζ) Orionis. It can be quite difficult to see, usually requiring a dark sky and at least an 8 inch (200 mm) telescope.

NGC 2169 This open cluster is made up of about 30 stars.

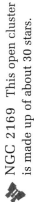

ORION NEBULA The Orion Nebula is one of the most impressive sights for skywatchers. It is pictured here with its companion, M 43 (the ball-like shape at top right).

Pavo

PAH-voh

The Peacock

Pavo, the Peacock, lies not far from the south celestial pole, south of Sagittarius and Corona Australis.

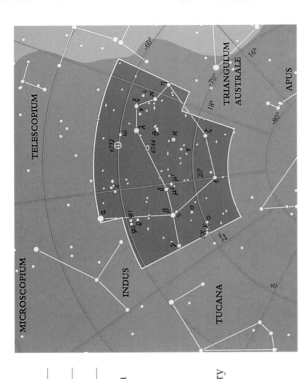

◈ THE PEACOCK STAR
(ALPHA [α] PAVONIS)   This star is 150 light-years away. It is a binary system whose members orbit each other in less than two weeks, but the pair is too close to separate telescopically.

✦ NGC 6752   A spectacular globular cluster at a relatively close distance of 17,000 light-years, this huge family of stars is the third largest globular cluster (in apparent size); only Omega (ω) Centauri and 47 Tucanae exceed it.

↗ NGC 6744   This faint but beautiful galaxy is one of the largest known barred spirals. Smaller telescopes reveal only the nuclear regions, a 10 inch (250 mm) telescope being necessary to reveal anything more.

PAVONIS (PAV)
On meridian
10 p.m. Aug 10

Pegasus

PEG-a-sus

The Winged Horse

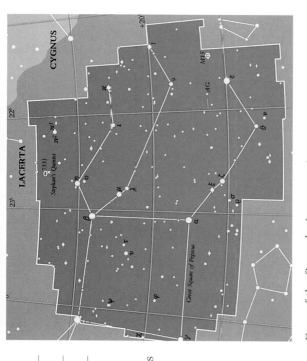

Although this constellation has no really bright stars, it is easy to spot because its three brightest stars, plus Alpha (α) Andromedae, form the Great Square of Pegasus.

M 15   This is one of the best of the northern sky globular clusters. Binoculars show it as a nebulous patch, but in a telescope it is a real showpiece.

NGC 7331   This spiral galaxy is the brightest one in Pegasus, but is still only 9th magnitude.

STEPHAN'S QUINTET   This very faint group of galaxies lies half a degree south of NGC 7331.

Four of the five galaxies appear to be interacting, distorting each other, and drawing out long streamers of stars. Stephan's Quintet is not really a target for the beginner, as it needs at least a 10 inch (250 mm) telescope to be seen clearly.

PEGASI (PEG)
On meridian
10 p.m. Oct 1

277

Perseus

PURR-see-us

The Hero

A pretty constellation that straddles the Milky Way, Perseus is in the northern skies from July to March.

 ALGOL   This is the most famous of the eclipsing variables. Every 2 days, 20 hours and 48 minutes, it begins to drop in brightness from magnitude 2.1 to 3.4 in an eclipse lasting 10 hours.

 M 34   This bright open cluster sits in the middle of a rich field of stars.

 THE DOUBLE CLUSTER (NGC 869 AND 884)   Two of the best examples of open clusters, NGC 869 and 884 (h Persei and Chi [χ]

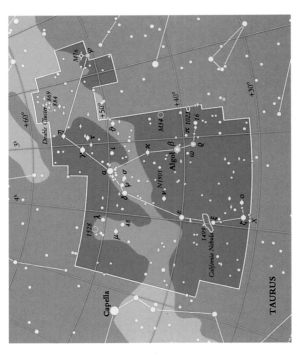

Persei respectively) are magnificent through binoculars or the low-power field of a small telescope.

 PERSEIDS   One of the best meteor showers, the Perseids peak on August 11 and 12.

PERSEI (PER)
On meridian
10 p.m. Dec 10

Phoenix

FEE-nicks

The Phoenix

A symbol of rebirth, in mythology the Phoenix was a bird of great beauty that lived for 500 years. It would then build a nest of twigs and fragrant leaves which would be lit by the noontime rays of the Sun. The Phoenix would be consumed in the fire, but a small worm would wriggle from the ashes, bask in the sun, and evolve into a new Phoenix.

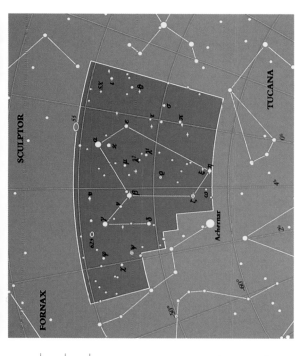

 SX PHOENICIS The best example of a "dwarf Cepheid" variable, this star changes from magnitude 7.1 to 7.5 and back again in only 79 minutes and 10 seconds! Cepheid periods are usually very exact. In this case, however, the range varies, with some maxima as bright as 6.7. The variation probably occurs because the star has two different oscillations occurring at once. Such a small range in brightness can be difficult to monitor, requiring careful comparison with neighboring stars.

PHOENICIS (PHE)
On meridian
10 p.m. Nov 1

Pictor

PIK-tor

The Painter's Easel

This southern constellation was originally named Equuleus Pictoris, the Painter's Easel, by Nicolas-Louis de Lacaille. Nowadays its shortened name refers solely to the painter. It is a dull group of stars lying south of Columba and alongside the brilliant star Canopus.

◆ BETA (β) PICTORIS This 4th-magnitude star is host to a disk of dust and ices that could be a planetary system in formation. The surrounding nebula is only visible using large telescopes.

KAPTEYN'S STAR Only 12.7 light-years away, this star was discovered by the Dutch astronomer

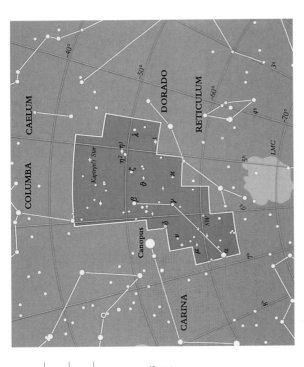

Jacobus Kapteyn in 1897. It moves quickly among distant background stars, crossing 8.7 arcseconds of sky per year—the width of the Moon every two centuries. At magnitude 8.8, the star is visible in binoculars and small telescopes.

Pisces

PIE-seez

The Fish

For thousands of years, this faint zodiacal constellation has been seen either as one fish or two. The ring of stars in the western fish, which is beneath Pegasus, is called the Circlet. The eastern fish is beneath Andromeda.

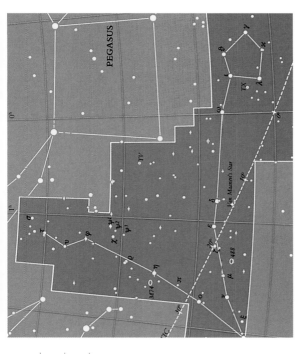

ZETA (ζ) PISCIUM A fine double star of magnitudes 5.6 and 6.5, separated by 24 arcseconds.

M 74 This is a large spiral galaxy, seen face-on, close to Eta (η) Piscium. While it is the brightest Pisces galaxy, it is still rather faint and requires a dark sky and an 8 inch (200 mm) telescope or larger to be seen. Photographs highlight its prominent nucleus and well-developed spiral arms.

VAN MAANEN'S STAR An 8 inch (200 mm) telescope will identify this magnitude 12.2 white dwarf star.

281

Piscis Austrinus

PIE-sis oss-STRINE-us

The Southern Fish

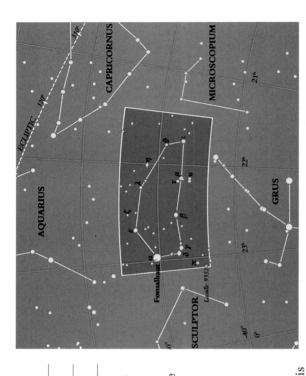

Lying to the south of Aquarius and Capricornus, Piscis Austrinus, the Southern Fish, is relatively easy to spot because of its lone, magnitude 1.2 star, Fomalhaut, which is often referred to as the Solitary One. For the Persians, 5,000 years ago, this was a Royal Star that had the privilege of being one of the guardians of heaven. Many early charts of the heavens show the Southern Fish drinking water that is being poured from Aquarius's jar.

◆ FOMALHAUT This bright star is 25.1 light-years away—close by stellar standards. It is about twice as large as our Sun and has 14 times its luminosity. Some 2 degrees of arc southward is a magnitude 6.5 dwarf star that seems to be sharing Fomalhaut's motion through space. They are so far apart that it is hard to call them a binary system. Maybe these two stars are all that is left of a cluster that dissipated long ago.

PISCIS AUSTRINUS
(PSA)
On meridian
10 p.m. Oct 1

Puppis and Pyxis

PUP-iss   PIK-sis

The Stern   The Compass

With the Milky Way running along it, Puppis provides a feast of open clusters for binoculars or telescopes. Right alongside is the smaller and fainter constellation of Pyxis.

◉ ZETA (ζ) PUPPIS This blue supergiant sun is one of our galaxy's largest. It shines at 2nd magnitude.

◉ L² PUPPIS One of the brightest of the red variable stars, L² Puppis varies from magnitude 2.6 to 6.2 over a period of five months.

✷ M 46 A beautiful open star cluster, through small telescopes, M 46 is a cloud of faint stars the apparent diameter of the Moon.

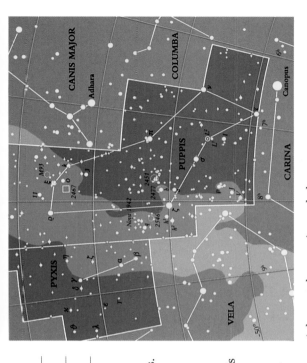

A larger telescope is needed to see NGC 2438, a planetary nebula superimposed on the cluster.

 T PYXIDIS This recurrent nova sometimes reaches 7th magnitude during its outbursts.

283

Reticulum

reh-TIK-u-lum

The Reticule

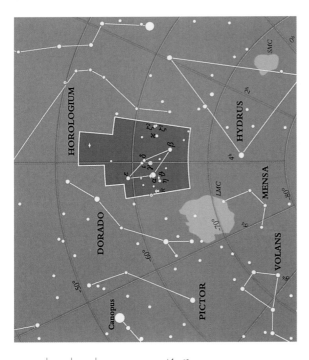

RETICULI (RETI)
On meridian
10 p.m. Dec 10

A small constellation of faint stars halfway between the bright stars Achernar and Canopus, Reticulum was first set up as Rhombus by Isaak Habrecht of Strasburg. Nicolas-Louis de Lacaille changed its name to Reticulum to honor the reticule—the grid of fine lines in a telescope eyepiece that aids with centering and focusing. It is occasionally also known as the Net.

 R RETICULI This Mira-type variable star is quite red, and at maximum light it shines at about magnitude 7. Over a period of nine months, R Reticuli drops to magnitude 13, then returns to maximum brightness.

284

Sagitta

sa-JIT-ah

The Arrow

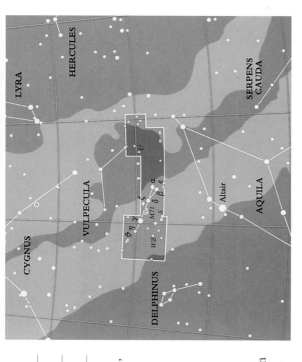

Although only a small constellation, Sagitta is easy to find halfway between Altair in Aquila, and Albireo (Beta [β] Cygni). The ancient Hebrews, Persians, Arabs, Greeks, and Romans all saw this group of stars as an arrow.

U SAGITTAE Every 3.4 days, this eclipsing binary drops from magnitude 6.5 to a minimum of 9.3.

V SAGITTAE Although this star is faint, varying erratically from magnitude 8.6 to magnitude 13.9, it is interesting because of the way it alters a little almost every night. It might have been a nova a long time ago.

 M 71 A little south of the midpoint of a line joining Delta (δ) and Gamma (γ) Sagittae, M 71 is a fertile cluster of faint stars. It is generally regarded as a poor globular cluster, rather than as a rich open cluster.

285

Sagittarius

sadge-ih-TAIR-ee-us

The Archer

This constellation is located in the Milky Way in the direction of the center of the galaxy. Here the band of the Milky Way is at its broadest, although cut by dark bands of dust. It is a treasure trove of galactic and globular clusters, plus bright and dark nebulas.

M 22 The Great Sagittarius star cluster is a very large globular. At magnitude 6.5, it is an easy object to see in binoculars, but a telescope really brings out the cluster's beauty. Only 10,000 light-years away, it is one of the closest globulars, and with an 8 inch (200 mm) telescope you should be able to resolve it into countless stars.

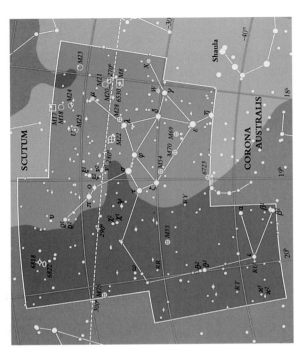

 M 23 Just one of many galactic clusters in Sagittarius, M 23 presents more than a hundred stars in an area about the size of the Moon. It is a striking sight in binoculars or a telescope using low magnification.

286

THE LAGOON NEBULA (M 8)

This spectacular diffuse nebula envelops the cluster of stars called NGC 6530. On a dark night the nebula is visible to the naked eye as a milky-white spot of light just to the north of one of the richest part of the Milky Way. In photographs, the extensive nebula is marked by several tiny dark splotches. Dutch astronomer Bart Bok identified these as globules in which new stars are being formed.

THE TRIFID NEBULA (M 20)

Found only 1½ degrees to the northwest of the Lagoon Nebula, the Trifid Nebula is likely to be part of the same complex of nebulosity. It is known as the Trifid because three lanes of dark clouds divide the nebula in the most beautiful way. You should be able to detect these dark lanes with a 6 inch (150 mm) telescope under a good sky.

THE OMEGA NEBULA (M 17)

Also called the Swan, the Horseshoe or the Check-mark, this nebula can be seen quite clearly in binoculars. In a 4 inch (100 mm) telescope it looks like the figure 2 with an extended baseline. It is a stunning sight in a larger telescope.

TRIFID NEBULA The photo clearly show the dust lanes which trisect M 20 and give it its popular name.

Scorpius

SKOR-pee-us

The Scorpion

A beautiful constellation of the zodiac, filled with bright stars and rich star fields of the Milky Way, Scorpius really does resemble a scorpion, complete with head and stinger. Near the northern end is a line of three bright stars, with red Antares at its center.

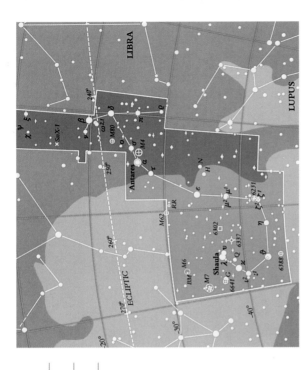

SCORPII (SCO)
On meridian
10 p.m. July 1

◉ ANTARES The Romans called this magnitude 1 star Cor Scorpionis, meaning "heart of the scorpion." Antares is a red super-giant 600 million miles (1 billion km) across and is 11,000 times more luminous than the Sun. However, with a mass only 10 or 15 times that of the Sun, it is not very dense. The star lies about 600 light-years away.

✂ BETA (β) SCORPII This is a double star whose 2.6 and 4.9 magnitude components are 13.7 arcseconds apart, making resolution possible even in a 2 inch (50 mm) telescope. Beta (β) Scorpii is also known as Graffias.

M 4 This strange globular cluster has a different appearance with each instrument you use. Binoculars show a fuzzy patch of light; a small telescope shows a large patch of mottled haze; and 4 or 6 inch (100 to 150 mm) instruments begin to show the individual stars. This is one of the best globulars for viewing in small telescopes.

THE BUTTERFLY CLUSTER (M 6) The stars of this bright open cluster really resemble a butterfly when viewed at high power.

M 7 This large, bright open cluster, lying to the southeast of M 6, needs to be seen through the large field of view of binoculars to be fully appreciated.

NGC 6231 Half a degree north of Zeta (ζ) Scorpii, this bright open cluster lies in a rich region of the Milky Way. It is best surveyed in binoculars or at very low power in a telescope.

M 80 This bright globular cluster can be seen in binoculars but needs a 10 inch (250 mm) telescope to resolve its stars.

SCORPIUS X-1 This is a close binary star in which one star expels gas onto a dense neighbor that could be either a white dwarf, a neutron star, or a black hole. It is a bright source of X-rays, but appears visually as a 13th-magnitude star.

THE SCORPION In Greek mythology, Scorpius is the tiny Scorpion that stung and killed Orion the Hunter.

289

Sculptor

SKULP-tor

The Sculptor

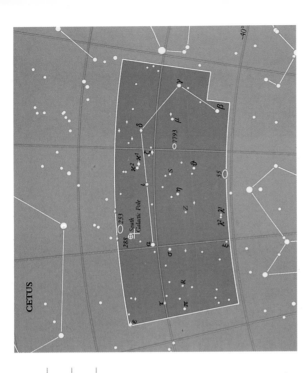

This constellation lies to the south of Aquarius and Cetus. Its most significant feature is a small cluster of nearby spiral galaxies.

NGC 253   For a small telescope user, this magnitude 7 galaxy is one of the most satisfying, especially for observers in the Southern Hemisphere. It is very large and is viewed almost edge-on. It was discovered by Caroline Herschel one night in 1783, while she was searching for comets. It appears as a thick streak in binoculars and begins to show the texture evident in photographs when larger instruments are used. The galaxy is 10 million light-years distant.

NGC 55   This is another very fine edge-on galaxy, similar to NGC 253, although fainter at 8th magnitude. Those using an 8 inch (200 mm) telescope or larger will see that it is distinctly brighter at one end than the other.

SCULPTORIS (SCL)
On meridian
10 p.m. Oct 20

Scutum

SKU-tum

The Shield

Although Scutum is not a large constellation and has no bright stars, it is not difficult to find in a dark sky because it is the home of one of the Milky Way's most dramatic clouds of stars. The constellation was created at the end of the 17th century, and named Scutum Sobiescianum (Sobieski's Shield) in honor of King John Sobieski of Poland.

 R SCUTI This semi-regular RV Tauri-type variable star changes from magnitude 5.7 to 8.4 and back over about five months.

 THE WILD DUCK CLUSTER (M 11) This spectacular open cluster is clearly visible in binoculars, rewarding in a small telescope, and stunning in an 8 inch (200 mm) one. One of the most compact of all the open clusters, the presence of a bright star in the foreground adds to its beauty.

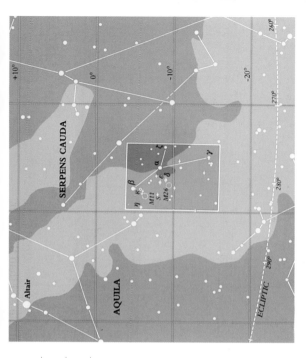

SCUTI (SCT)
On meridian
10 p.m. Aug 1

Serpens

SIR-penz

The Serpent

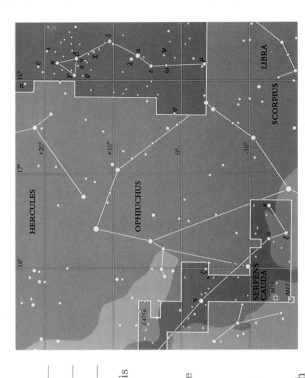

SERPENTIS (SER)
On meridian
10 p.m. June 20
(Serpens Caput)
10 p.m. July 20
(Serpens Cauda)

This is the only constellation that is divided into two parts. The head (Serpens Caput) and the tail (Serpens Cauda) are separated by the constellation of Ophiuchus, the Serpent Bearer.

R SERPENTIS A Mira star almost midway between Beta (β) and Gamma (γ) Serpentis, this variable has a bright maximum of 6.9. It fades to about 13.4, although it sometimes can become fainter. Its period is about one year.

M 5 This very striking 5th-magnitude globular cluster in Serpens Cauda is about 26,000 light-years away.

THE EAGLE NEBULA (M 16) Through an 8 inch (200 mm) or larger telescope on a dark night, this combination of nebula and star cluster is stunning. But you can still enjoy the sight of the cluster in smaller telescopes.

Sextans

SEX-tanz

The Sextant

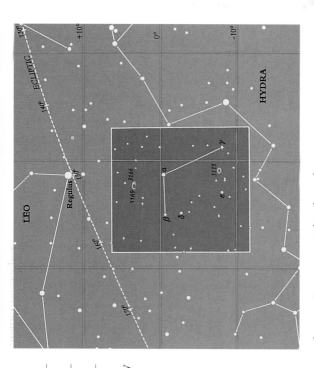

Sextans Uraniae, now known simply as Sextans, was the creation of the 17th-century German astronomer Johannes Hevelius. He chose this name for the new constellation to commemorate the loss of the sextant he once used to measure the positions of the stars. Along with all his other astronomical instruments, the sextant was destroyed in a fire that took place in September 1679. "Vulcan overcame Urania," Hevelius remarked sadly, commenting on the fire god having defeated astronomy's muse.

THE SPINDLE GALAXY (NGC 3115) Because we see this 10th-magnitude galaxy almost edge-on, it appears to be shaped like a lens, with a bright center. Unlike many faint galaxies, the Spindle Galaxy gives quite satisfying views at high power. It seems to be somewhere between an elliptical and a spiral.

293

Taurus

TORR-us

The Bull

Taurus is a prominent constellation clearly visible from both Northern and Southern Hemispheres, from February through April.

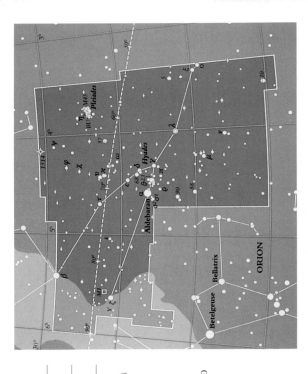

👁 THE PLEIADES (M 45) Also known as the Seven Sisters, this is the most famous open star cluster in the sky and forms the bull's shoulder. Alcyone (Eta [η] Tauri) is the most dazzling sister. She is accompanied by: Maia (20 Tauri); Asterope I and II (the double star 21 Tauri); Taygeta (19 Tauri); Celaeno (16 Tauri); and Electra (17 Tauri). Finally, there is Merope (23 Tauri), a star surrounded by a beautiful cloud of cosmic grains producing a blue reflection nebula. On a reasonably dark night, you should be able to see at least six of the stars in the Pleiades with the naked eye; under good conditions, you might be able to see as many as nine. Containing more than 500 stars, the Pleiades is about 400 light-years

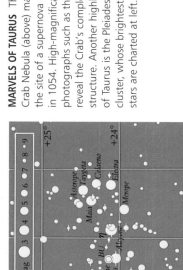

MARVELS OF TAURUS The Crab Nebula (above) marks the site of a supernova seen in 1054. High-magnification photographs such as this reveal the Crab's complex structure. Another highlight of Taurus is the Pleiades cluster, whose brightest stars are charted at left.

away and covers an area four times the size of the Full Moon. It is best seen with binoculars.

THE HYADES Like the Pleiades, this is also an open cluster, but it is so close to us (only 150 light years away) that even when viewed with the naked eye the stars appear to be spread out. The stars of the Hyades form the bull's head.

ALDEBARAN (ALPHA [α] TAURI) This is an orange giant and is the brightest star in Taurus. Only 60 light-years away, it marks the bull's eye.

THE CRAB NEBULA (M 1) This nebula is clearly visible in a 4 inch (100 mm) telescope on a dark night as an oval glow, 5 arcminutes across. Details in the cloud can be detected in 10 inch (250 mm) scopes or larger.

Telescopium

tel-eh-SKO-pee-um

The Telescope

Originally bearing the name Tubus Telescopium, this constellation was created by Nicolas-Louis de Lacaille during the 18th century to honor the invention of the telescope. Lacaille "borrowed" stars from large surrounding constellations in order to create his new one.

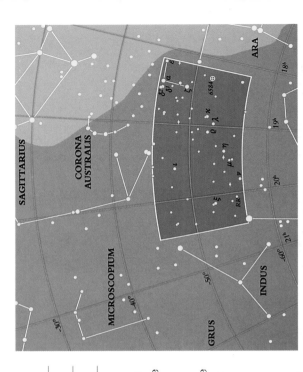

TELESCOPII (TEL)
On meridian
10 p.m. Aug 10

RR TELESCOPII Although this star is normally too faint for small telescopes, it is one of the most interesting novas on record. Before 1944, this star varied over about 13 months between 12.5 and 15th magnitude, but in that year it began a rise to magnitude 6.5 that took some five years. As the nova declined in the following years, it still displayed its original 13-month period. It is thought that the star may be a binary system, in which a large red star is responsible for the minor variations that take place, and a smaller, hotter star puts on the nova part of the performance.

296

Triangulum

tri-ANG-gyu-lum

The Triangle

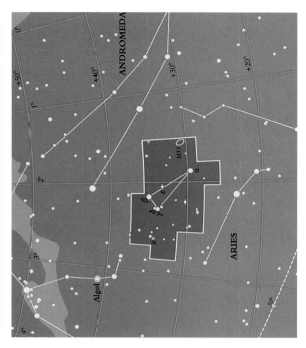

TRIANGULI (TRI)
On meridian
10 p.m. Nov 20

This is a small, faint constellation extending just south of Andromeda, near Beta (β) and Gamma (γ) Andromedae. The group of stars was known to the ancients, and because of its similarity to the Greek letter delta (Δ) it was sometimes called Delta or Deltoton. The ancient Hebrews gave it the name of a triangular musical instrument.

THE PINWHEEL GALAXY

(M 33) This is one of the brightest members of our Local Group and we have a front-row view because it appears face-on. It shines at magnitude 5.5 but its light is spread out over such a large area that it is difficult to see. Although it can be seen by the naked eye on very clear nights, you need a dark sky and binoculars to see a fuzzy glow larger than the Full Moon. A telescope with a wide field of view will also show the galaxy, but one with a narrow field will show nothing at all.

297

Triangulum Australe

tri-ANG-gyu-lum os-TRAH-lee

The Southern Triangle

Triangulum lies just to the south of Norma, the Level, and to the east of Circinus, the Drawing Compass—tools used by woodworkers and navigators on early expeditions to the Southern Hemisphere.

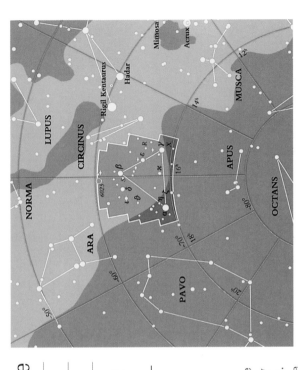

TRIANGULI
AUSTRALIS (TRA)
On meridian
10 p.m. June 20

R TRIANGULI AUSTRALIS One of several Cepheids in the constellation, this variable alters by about a magnitude—from 6.0 to 6.8. Because it is a Cepheid variable, we know its period precisely, which is 3.389 days. For Cepheids with this rapid variation, magnitude estimates at least once a night are worthwhile.

S TRIANGULI AUSTRALIS Another Cepheid variable, S

varies from magnitude 6.1 to 6.7 and back over a period of 6.323 days.

NGC 6025 This is a small open cluster of about 30 stars of 9th magnitude, with fainter background stars.

Tucana

too-KAN-ah

The Toucan

Johann Bayer first published this constellation in his star atlas of 1603. From the earliest drawings, the Tucan sat on the Small Magellanic Cloud, one of the two closest galaxies to the Milky Way, tending it like an egg.

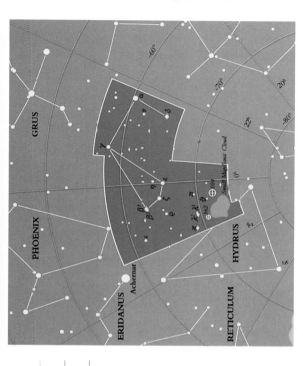

◐ 47 TUCANAE (NGC 104) From its perch 16,000 light-years away, this glorious globular cluster shines brightly at magnitude 4.5. Although it is a naked-eye object under dark skies, a 4 inch (100 mm) or larger telescope really brings out the best in this cluster, which competes with Omega (ω) Centauri for the title of the most splendid globular cluster in the entire sky.

◐ THE SMALL MAGELLANIC CLOUD (SMC) This galaxy is visible to the naked eye on a good night, with the great globular cluster 47 Tuc alongside. About 192,000 light-years away, the cloud is some 30,000 light-years wide.

299

Ursa Major

ER-suh MAY-jer

The Great Bear

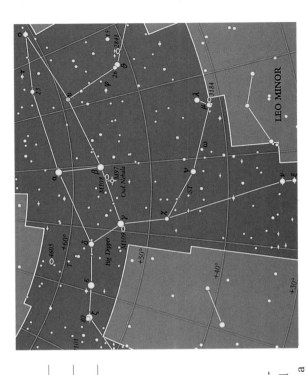

URSAE MAJORIS
(UMA)
On meridian
10 p.m. April 20

This well-known constellation contains the group of seven stars that make up the Big Dipper. In Greek myth, Zeus and Callisto, a mortal, had a son called Arcas. Hera, Zeus's jealous wife, turned Callisto into a bear, and one day while out hunting, her son, not knowing that the bear was his mother, almost killed her. Zeus rescued Callisto, placing both her and her son, whom he also turned into a bear, in the sky together. Callisto is Ursa Major and Arcas is Ursa Minor.

◉ MIZAR (ZETA [ζ] URSAE MAJORIS) AND ALCOR Mizar and Alcor make up the famous apparent double star in the middle of the Big Dipper's handle. The two stars are separated by 12 arcminutes and it is thus possible to see them as a pair with the naked eye, although your eyesight needs to be sharp. Mizar is itself a true binary star, its components separated by 14 arcseconds.

300

M 81 This spiral galaxy can be easily seen through binoculars, even when observing in the city, and it is dramatic when observed under good conditions. The oval disk becomes more apparent with increasing telescope size. M 81 is probably a fair representation of how the Milky Way Galaxy would look from the outside.

M 82 This is a long, thin, peculiar galaxy, half a degree from M 81. It appears as a thin, gray nebulosity in a 4 inch (100 mm) telescope, but begins to show some detail in an 8 inch (200 mm) or larger one. Even in large telescopes or photographs, however, it is not clear what type of galaxy this is.

M 101 This large, spread-out spiral galaxy is visible through small telescopes if the sky is dark enough. It needs a wide field and

a low-power eyepiece. At 16 million light-years, it is one of the closer spiral galaxies to the Milky Way.

The Owl Nebula (M 97) This is an oval planetary nebula that takes the shape of an owl when it is seen in a 12 inch (300 mm) telescope. It is large and dim, and a 3 inch (75 mm) or larger telescope is needed to find it.

ANCIENT BEAR Ursa Major is one of the oldest constellations, and many civilizations depicted it as a bear. The ancient Egyptians, however, considered this group of stars to be either a hippo or a Nile River boat for the god Osiris.

Ursa Minor

ER-suh MY-ner

The Little Bear

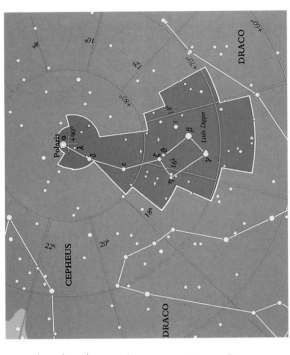

Ursa Minor looks a bit like a spoon whose handle has been bent back by a playful child. This group of stars was recognized as a constellation in 600 BC by the Greek astronomer Thales. The Little Bear, according to Greek legend, is Arcas, son of Callisto—Ursa Major, the Great Bear. Placed in the heavens by Zeus, he and his mother follow each other around the north celestial pole.

◆ POLARIS (ALPHA [α] URSAE MINORIS) The pole star for the Northern Hemisphere, this Cepheid variable is currently almost 1 degree from the exact pole. Precession of the Earth's axis will carry the pole to within about 27 arcminutes of Polaris around the year 2100, and then it will start to move away again. Polaris is 820 light-years away, with a 9th magnitude companion some 18½ arcseconds away. Splitting this pair is an interesting test for a 3 inch (75 mm) telescope.

Vela

VEE-lah

The Sail

This constellation was created in the 1750s and represents the sail of the ship *Argo*, in which Jason and the Argonauts sailed to search for the Golden Fleece.

 THE FALSE CROSS Delta (δ) and Kappa [κ] Velorum, together with Epsilon (ε) and Iota (ι) Carinae, make up a larger but fainter version of the Southern Cross which is known as the False Cross.

 GAMMA (γ) VELORUM This double star is resolvable in a steady pair of binoculars.

 NGC 3132 This bright planetary nebula accompanies the many clusters in Vela, but lies right on the border with Antlia. Being 8th magnitude and almost 1 arcminute across, it is considered the southern version of Lyra's Ring Nebula, but with a much brighter central star.

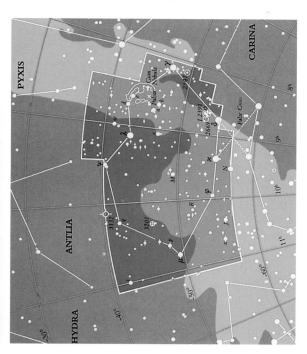

VELORUM (VEL)
On meridian
10 p.m. March 10

303

Virgo

VER-go

The Maiden; The Virgin

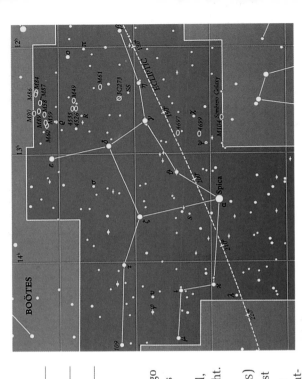

Scattered throughout Virgo and Coma Berenices are more than 13,000 galaxies. Known as the Virgo Cluster or Coma-Virgo Cluster, this mighty club of distant systems of stars repays sweeping with a small, wide-field telescope on a dark night.

👁 SPICA (ALPHA [α] VIRGINIS) A bright, white star, Spica is almost exactly 1st magnitude, although it has a slight variation. It is 220 light-years away and more than 2,000 times as luminous as the Sun.

🔭 PORRIMA (GAMMA [γ] VIRGINIS) This is one of the best double stars in the sky, each component shining at magnitude 3.7. At 3 arcseconds separation, the pair is easy to separate.

🔭 M 84 AND 86 These two elliptical galaxies are close enough to be seen in the same low-power telescope field. On a dark

night, an 8 inch (200 mm) telescope will show several smaller galaxies in the same view.

M 87 This elliptical galaxy is one of the mightiest galaxies we know. Through a small telescope it appears as a bright patch of fuzzy light about a magnitude brighter than M 84 and 86. Interestingly, larger telescopes do not show a great deal more. In the professional size range, however, more details do emerge. With a 60 inch (1.5 m) telescope, for example, you can see a jet emerging from the galaxy's center.

THE SOMBRERO GALAXY (M 104) Although M 104 is quite a distance south of the main concentration of galaxies, it seems to be gravitationally attracted to the swarm and so is thought to be a part of it. The brightest of the Virgo galaxies, a dark lane cuts along its equator, making it look a little like a sombrero hat in an 8 inch (200 mm) telescope.

3C273 VIRGINIS This is the brightest known quasar, but being only 13th magnitude, an 8 inch (200 mm) scope is needed.

SOMBRERO GALAXY
The Sombrero Galaxy presents a bright, 8th-magnitude glow, just 8 arcminutes across, but is readily seen in smaller telescopes.

Volans

VOH-lanz

The Flying Fish

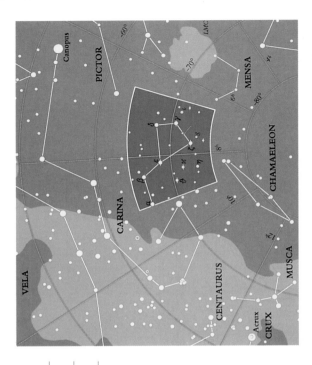

VOLANTIS (VOL)
On meridian
10 p.m. Feb 20

Piscis Volans, the Flying Fish, lies south of Canopus, and was introduced in 1603. It is now known only as Volans. Sailors in the south seas had reported seeing schools of flying fish, which may have been the inspiration for the name. The pectoral fins of these fish are as large as the wings of birds and they glide across the water for distances of up to a quarter of a mile (400 m).

S VOLANTIS   A Mira star,
S Volantis usually has a maximum magnitude of 8.6, but it has occasionally risen to 7.7. Its faint minimum averages 13.6. The star completes its cycle in a little less than 14 months.

Vulpecula

vul-PECK-you-lah

The Fox

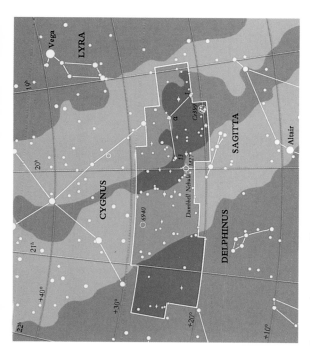

VULPECULA (VUL)
On meridian
10 p.m. Aug 20

This constellation, invented by Johannes Hevelius in 1690, is without an exciting story or a moral tale. Hevelius's name for it was Vulpecula cum Anser, the Fox with the Goose, but now the constellation is simply referred to as the Fox.

 THE DUMBBELL NEBULA (M 27) This is one of the finest planetary nebulas in the sky and is well suited to small telescopes. Bright and large, it is easy to find just north of Gamma (γ) Sagitlae. Being 7th magnitude, it can be found through binoculars, but it appears only as a faint nebulous spot. If you use a small telescope, you can make out its odd shape. A larger telescope will reveal its 13th-magnitude central star. Although the nebula's gases are expanding at the rate of 17 miles (27 km) per second, there will be no noticeable change in the appearance of the nebula within a human lifetime.

307

GLOSSARY

Active galaxy A galaxy with
a central black hole that is
emitting lots of radiation.

Aperture The diameter of
a telescope's main light-collecting
optics. Also, the diameter of a
binocular lens.

Arcminute A unit of angular
measure equal to 1/60 of a degree;
the Moon and Sun are about
30 arcminutes across.

Arcsecond A unit of angular
measure equal to 1/60 of an
arcminute; Jupiter averages some
44 arcseconds across.

Astronomical unit (AU) The
average distance between Earth
and the Sun, about 93 million
miles (150 million km).

Atmosphere The layer of gases
attached to a planet or moon
by gravity.

Axis The imaginary line through
the center of a planet, star, or
galaxy around which it rotates;
also, a shaft around which a
telescope mounting pivots.

Big Bang The explosion of a
small, very hot lump of matter
about 15 billion years ago that
marked the birth of the universe,
according to the current theory of
the universe's origin.

Binary star (double star) Two
stars linked by mutual gravity
and revolving around a common
center of mass.

Black hole An object so dense
that no light or other radiation
can escape from inside it.

Celestial equator The imaginary
line encircling the sky midway
between the celestial poles.

Celestial poles The imaginary
points on the sky where Earth's
rotation axis, extended infinitely,
would touch the imaginary
celestial sphere.

Celestial sphere The imaginary
sphere enveloping Earth upon

which the stars, galaxies, and other celestial objects all appear to lie.

Coma A comets' head, consisting of a cloud of dust and gas.

Comet A small body composed of ice and dust that orbits the Sun on an elongated path.

Constellation One of the 88 official patterns of stars that divide the sky into sections.

Declination The angular distance of a celestial object north or south of the celestial equator.

Degree A unit of angular measure equal to $\frac{1}{360}$ of a circle.

Electromagnetic spectrum The full range of radiation produced by nature, from gamma rays to radio waves.

Focal length The distance between the main lens or mirror of a telescope and the point where the light from it comes to a focus.

Galaxy A huge gathering of stars, gas, and dust, bound by gravity and having a mass ranging from 100,000 to 10 trillion times that of the Sun.

Gamma rays Radiation with a wavelength shorter than X-rays.

Gas-giant planet A planet whose composition is dominated by hydrogen: Jupiter, Saturn, Uranus, and Neptune.

Globular star cluster A spherical cluster that may contain up to a million stars.

Infrared (IR) Radiation with wavelengths just longer than those of visible light.

Light-year The distance that light travels in one year, about 6 trillion miles (9.5 trillion km).

M objects Star clusters, nebulas, and galaxies in the Messier list.

Magnitude A unit of brightness for celestial objects. Apparent magnitude describes how bright a star looks from Earth, while absolute magnitude is its brightness if placed at a distance of 32.6 light-years.

Meteor A meteoroid (a piece of space debris) that has entered Earth's atmosphere at high speed and begun to burn up, producing a bright, transient streak of light.

Meteorite Any piece of inter-planetary debris that reaches Earth's surface intact.

Microwave Radiation with wavelengths measured in millimeters.

Nebula A cloud of gas or dust in space; may be either dark or luminous.

NGC objects Galaxies, star clusters, and nebulas listed in the New General Catalogue.

Nova A white dwarf star in a binary system that brightens suddenly by several magnitudes as gas pulled away from its companion star explodes in a thermonuclear reaction.

Nucleus The central core of a galaxy or comet.

Open star cluster A group of a few hundred stars bound by gravity and moving through space together.

Orbit The path of an object as it moves through space under the control of another's gravity.

Planetary nebula A shell of gas puffed off by a star late in its life.

Pulsar An old, rapidly spinning star that flashes bursts of radio (and occasionally optical) energy.

Quasar Short for quasi-stellar radio source, quasars are thought to be the active nuclei of very distant galaxies.

Radiant The point on the sky from where a shower of meteors appears to come.

Radiation The means by which energy travels through space.

Radio Radiation of centimeter or longer wavelength.

Radio astronomy The study of celestial bodies by means of the radio waves that they emit and absorb.

Red giant A large, cool, red star in a late stage of its life.

Right ascension (RA) The celestial coordinate analogous to longitude on Earth.

Seeing A measure of the steadiness of the atmosphere. Good seeing is essential to using high magnifications.

Sunspot A dark region on the Sun's surface, cooler than the surrounding area.

Supernova The explosion of a massive star in which it blows off its outer atmosphere and briefly equals a galaxy in brightness.

Supernova remnant An expanding cloud of gas that has been thrown into space by a supernova explosion.

Terrestrial planet A planet whose composition is mainly rocky: Mercury, Venus, Earth, and Mars.

Ultraviolet (UV) Radiation with wavelengths just shorter than those of visible light.

Variable star Any star whose brightness appears to change; periods may range from minutes to years.

Wavelength The distance between two successive waves of energy passing through space.

White dwarf The small, very hot remnant of a star that has evolved past the red giant stage.

X-rays Radiation with wavelengths between ultraviolet and gamma rays.

INDEX

Page numbers in *italics* indicate illustrations and photos